PENGUIN MODERN MASTERS
EINSTEIN

Jeremy Bernstein is a physicist who frequently writes on scientific subjects for nonscientists. Educated at Harvard and currently professor of physics at Stevens Institute of Technology, he has held visiting professorships at Oxford, Rockefeller University, the University of Islamabad in Pakistan, and CERN in Geneva. Books to his credit include *A Comprehensible World, The Wildest Dreams of Kew, Ascent,* and *The Analytical Engine.*

Frank Kermode, King Edward VII Professor of English Literature at Cambridge, is the author of *The Classic; D.H. Lawrence; Shakespeare, Spenser, Donne;* and other widely acclaimed critical studies.

MODERN MASTERS

ALREADY PUBLISHED

PENGUIN
MODERN MASTERS

EDITED BY frank kermode

*By Modern Masters we mean the men who
have changed and are changing the life and
thought of our age. The authors of these volumes are themselves masters, and they have
written their books in the belief that general
discussion of their subjects will henceforth be
more informed and more exciting than ever
before.* —F.K.

einstein

jeremy bernstein

PENGUIN BOOKS

Penguin Books Ltd, Harmondsworth,
Middlesex, England
Penguin Books, 625 Madison Avenue,
New York, New York 10022, U.S.A.
Penguin Books Australia Ltd, Ringwood, Victoria, Australia
Penguin Books Canada Limited, 2801 John Street,
Markham, Ontario, Canada L3R 1B4
Penguin Books (N.Z.) Ltd, 182–190 Wairau Road,
Auckland 10, New Zealand

First published in the United States of America by The Viking Press 1973
Published in Great Britain by Fontana 1973
Viking paperback edition published 1973
Reprinted 1974
Published in Penguin Books 1976
Reprinted 1978, 1979, 1980, 1982

LIBRARY OF CONGRESS CATALOGING IN PUBLICATION DATA
Bernstein, Jeremy, 1924—
Einstein. (Penguin modern masters)
Reprint of the ed. published by The Viking Press, New York,
in series Modern masters.
"Most of the material . . . appeared originally in the New Yorker."
Bibliography: p. 231. Includes index.
1. Einstein, Albert, 1879–1955.
[QC16.E5B45   1976]   530'.092'4   [B]   76-22629
ISBN 0 14 00.4317 9

Printed in the United States of America by
Kingsport Press, Inc., Kingsport, Tennessee
Set in Linotype Primer

Acknowledgment is made to the following for permission to publish excerpts from the works cited.
Cambridge University Press: From Space, Time and Gravitation
by Sir Arthur Eddington.
Daedalus: From Mach, Einstein, and the Search for Relativity.
Reprinted by permission of Daedalus, Journal of the American
Academy of Arts and Sciences, Boston, Mass., Spring 1968, His-
torical Population Studies.
Estate of Albert Einstein for all excerpts from Albert Einstein's
writings. Copyright © by Estate of Albert Einstein; for excerpts
from Albert Einstein: Philosopher-Scientist. Copyright © by Library
of Living Philosophers.
Alfred A. Knopf, Inc.: From Einstein and His Times, by Philipp
Frank. Copyright 1947, 1953 by Alfred A. Knopf, Inc. Reprinted by
permission of the publisher.
North-Holland Publishing Company: From Paul Ehrenfest by Martin J. Klein.
The Royal Society: From Lord Keynes, "Newton the Man," reprinted in The World of the Atom, edited by A. Boorse and Lloyd
Motz.

'I don't know what you mean,' said Alice.

'Of course you don't!' the Hatter said, tossing his head contemptuously.

'I dare say you never even spoke to Time!'

'Perhaps not,' Alice cautiously replied: 'but I know I have to beat time when I learn music.'

'Ah! that accounts for it,' said the Hatter. 'He won't stand beating. Now, if you only kept on good terms with him, he'd do almost anything you liked with the clock. For instance, suppose it were nine o'clock in the morning, just time to begin lessons: you'd only have to whisper a hint to Time, and round goes the clock in a twinkling! Half-past one, time for dinner!'

('I only wish it was,' the March Hare said to itself in a whisper.)

LEWIS CARROLL, *Alice's Adventures in Wonderland* 1865

I do not believe that there is any man now living who can assert with truth that he can conceive of time which is a function of velocity or is willing to go to the stake for the conviction that his 'now' is another man's future or still another man's 'past.'

W. F. MAGIE, *Presidential Address to the American Association for the Advancement of Science, 1911*

Thence we conclude that a balance clock at the equator must go more slowly, by a very small amount, than a precisely similar clock situated at one of the poles under otherwise identical conditions.

ALBERT EINSTEIN, *On The Electrodynamics of Moving Bodies*, 1905

PREFACE AND
ACKNOWLEDGMENTS

While it has become popular, at the present time, to deplore the role of science and technology in modern life, there has occurred, nonetheless—if the publication of new books is taken as evidence—a renewal of interest in the life and personality of Albert Einstein. Insofar as this interest extends to an understanding of what it was that Einstein added to modern thought, it is, I think, a hopeful sign. For it is very likely that future generations will refer to the first half of the twentieth century as the age of Einstein, just as historians think of the latter half of the seventeenth century as the age of Newton. In Newton's case we are far enough past the event to see how his thought permeated every aspect of the intellectual life of his successors—from the arts to philosophy and politics—for a hundred years. Not quite a quarter of a century has passed since Einstein's death in 1955, so it is as yet too early to judge the full impact of his work, although one may have intuitive judgments as to what this will be. The irony of this is that Einstein's work is understood by such a small percentage of the people whose lives and intellectual outlook have been, often unwittingly, influenced by it.

It is impossible in a short study such as this one or, in fact, in any popular study to convey the full richness of Einstein's contributions to modern science and modern thought. One may hope to convey at least the sense of it, and a sense of the man, and thereby arouse the reader's curiosity to explore deeper or, at least, to give the reader a feeling of kinship with those professional scientists who are devoting their lives to this exploration. With this in mind I have organized this book in a somewhat unconventional way. Rather than tracing Einstein's life and work chronologically I have organized the book around the three basic themes in his work: the special theory of relativity, the general theory of relativity and gravitation, and the quantum theory. Interpolated within the three basic sections of the book the reader will find what I hope is a consistent biographical study. My hope is that this organization will bring the reader deeper and deeper into Einstein's work, and deeper and deeper into the workings of the man insofar as I am able to understand them myself.

I am indebted to many people for their aid in preparing this book. Above all I am indebted to Professor Philipp Frank, who died in 1966 at the age of eighty-two. As the reader will see, there are many references in what follows to his book *Einstein: His Life and Times* and a good deal of anecdotal material which he told me during the 1950s, when I saw him frequently. My great regret is that, for most of this period, I was too ignorant to have asked him the numerous questions that occurred to me while writing this book. In addition I would like to thank my colleagues in physics Professors J. L. Anderson, M. A. B. Beg, S. Chandrasekhar, F. J. Dyson, G. Feinberg, G. Holton, M. J. Klein, A. Miller, A. Pais, V. Singh, and J. C. Taylor for reading and criticizing portions of the manuscript or for supplying additional material and encouragement. I am also indebted to Miss Helen Dukas for reading the manuscript and for a visit to Einstein's house in Princeton; and to Mrs. Elisabeth Sifton of The Viking Press for helpful editorial advice.

J. B.

CONTENTS

1. The Theory
of Relativity

Preamble: Einstein and the Queen of Belgium

The friendship of Albert Einstein with Albert and Elizabeth, the King and Queen of Belgium, which began in 1927 and lasted until Einstein's death in 1955, was made possible by the existence of the Solvay Congresses, which were held in Brussels. Ernest Solvay, a Belgian industrial chemist who had made a fortune with a new process of manufacturing sodium carbonate, dabbled in physics as a hobby and in 1911 conceived the notion of assembling the leading European physicists together, at his expense, and getting their opinions of his ideas. He left the organization of the congress to his friend Walther Nernst, then a professor in Berlin and a leading physical chemist. By the time the invitations to the first Solvay Congress of 1911 were issued, Nernst had succeeded in broadening the scope of the congress to a discussion of the central problems of physics. The Congress of 1911, which included such people as H. A. Lorentz, Max Planck, Madame Curie, Ernest Rutherford, as well as Einstein, was a great success, and it developed into an institution which lasted, on a periodic basis, until the present day.

It was these meetings that gave Einstein the chance to broaden his friendship with the royal couple, whom he referred to as "the Kings," as if this were a family name, and by the 1930 Solvay Congress their friendship was sufficiently close so that he simply dropped by to see them when he had some free time. As he wrote to his wife, Elsa,

> I went across to the station . . . to telephone the Kings. It was quite tedious because the line was always busy. . . . At 3 o'clock I drove out to the Kings, where I was received with touching warmth. These people are of a purity and kindness seldom found. First we talked for about an hour. Then an English woman musician arrived, and we played quartets and trios (a musical lady-in-waiting was also present). This went on merrily for several hours. Then they all went away and I stayed behind alone for dinner with the Kings—vegetarian style, no servants. Spinach with hard-boiled eggs and potatoes, period. (It had not been anticipated that I would stay.) I liked it very much there, and I am certain the feeling is mutual.[1]

By the spring of 1933 Einstein had taken refuge in the Belgian sea resort of Le Coq-sur-Mer. (He had actually been teaching the previous winter at the California Institute of Technology, and after Hitler came to power in January 1933 had decided that he could not return to his home in Berlin.) His summer home near Berlin had been sacked by the Gestapo and his property confiscated on the pretense that it was going to

[1] Otto Nathan and Heinz Norden, eds., *Einstein on Peace*, p. 661.

be used to finance a "Communist revolt." Some of his writings on the relativity theory had been burned publicly in the square before the State Opera House in Berlin, and Einstein had resigned from the Prussian Academy of Science to which he had been elected in 1913. In fact there were rumors of an assassination plot against Einstein's life, and these were taken seriously enough so that the Belgian government had provided him with bodyguards. (Philipp Frank, Einstein's successor in Prague, having heard on returning from a trip to London that Einstein was in Belgium, managed to locate him by asking around Le Coq even though the natives had been given strict orders not to disclose Einstein's whereabouts to anyone. "Einstein himself," Professor Frank wrote later, "laughed heartily at the failure of the measures taken by the police for his protection."[2]

By the late summer of 1933 Einstein had left Belgium for the United States, where he arrived on October 17. He was never to return to Europe. He had accepted a position at the newly founded Institute for Advanced Study in Princeton—in fact he was its first professor. It had been the idea of Abraham Flexner, the Institute's first director, to create a center at which young scholars could have informal contact with a small nucleus of first-rate permanent members. There were to be no rigid classroom schedules and the Institute would issue no degrees. (Flexner was also concerned that the permanent members should be as free as possible from financial worry. Einstein thought that $3000 a year would be a reasonable salary, which Flexner amended to $16,000.) Flexner also decided that the first members should be in mathematics and mathematical physics, since this work required little expensive technical support—a small library at most—and the disciplines were fundamental, with the feature that almost everyone

[2] Philipp Frank, *Einstein: His Life and Times*, p. 241.

could agree on who were the really outstanding people. The founding of the Institute, in 1930, coincided with the rise of Hitler, and this no doubt, helped enable Flexner to recruit not only Einstein but also some of the greatest European mathematicians, such as Hermann Weyl and John von Neumann—along with the American mathematician Oswald Veblen.

In 1933 Einstein moved into a rented house in Princeton, three quarters of a mile or so from what is now Fuld Hall at the Institute, where he had his office from 1940. In 1935 he bought a modest white frame house at 112 Mercer Street, where he lived and worked until his death. (The house, which now belongs to Margot Einstein, Einstein's stepdaughter, is shared by Helen Dukas, Einstein's secretary since 1928. Elsa Einstein, Margot's mother and Einstein's second wife, died in 1936.) The house was arranged so that Einstein had a small apartment on the second floor where he could shut himself off and work. The apartment is still much the way it was during his lifetime except that now, according to Miss Dukas, there are a few more plants. His study, which communicates directly to the bedroom, has a few rather stiff-backed chairs in it, and the table on which he worked. There is a large picture window facing the garden. The walls are lined with bookshelves, some of which house his record collection. Einstein was not an omnivorous reader. In literature his tastes ran mainly to the Russian novelists, like Dostoevski and Tolstoi. He was a great admirer of Gandhi and read aloud to his family Gandhi's autobiography as well as Herodotus and Frazer's *Golden Bough*. Most of the books, in every conceivable language, were sent to him by people who wanted him to read them. He almost never read any book that discussed himself. There are a few pictures and etchings on the walls—a drawing of Gandhi, photographs of his mother and his only sister, Maja, who moved to Princeton from Italy in 1939 and who lived with him until her death in 1951. He brought with him from Europe etchings of three of the physicists whom he admired most—Michael Faraday, James Clerk Maxwell, and, of course, Newton. The Faraday and Maxwell etchings are still there, but Newton has come out of his frame and has been replaced by a bit of abstract modern art. These, added to the utter simplicity of the rest, give the apartment a feeling of both serenity and solitude. "He did not like frills," as Miss Dukas has remarked.

When Einstein arrived in Princeton in 1933 he was fifty-four, and while he continued to do extensive work in physics it is fair to say that his great work, the work that created modern physics, was behind him. As he grew older he became more solitary both in his personal life and in the path in physics that he chose to follow. His external life style hardly varied from year to year. During the academic year—October to April—he went to his study at the Institute every day, even after his formal retirement in 1945, for a few hours. He never owned an automobile—there is not even a driveway at 112 Mercer Street—and even in his seventies he walked at least one way to the Institute (the Institute bus took him the other way), whatever the weather. During the summer months he went often to the sea, where he could indulge in his fondness for sailboats. He had various young assistants, but his contact with the physics community at the Institute and the University, as well as with the Princeton community as a whole, was very limited. Mostly this was by choice. Einstein never fully belonged to any institution, country, or person, even members of his family. He found great pleasure in the company of selected individuals, and he carried on an enormous and constant correspondence with people of all sorts, but there was always a sense about him that his thoughts, even his being, were elsewhere. He was fully aware of this and of the paradox of both his absolute need for solitude and the loneliness it produced. He was also aware, obviously, of his celebrity, which

he could sense but which he neither wanted nor understood. As he wrote to a friend in September 1952, "For
my part, I have always tended to solitude, a trait that
usually becomes more pronounced with age. It is
strange to be known so universally and yet to be so
lonely. The fact is that the kind of popularity which I
am experiencing pushes the subject into a defensive
attitude that leads him into isolation."[3]

During the twenty-two years Einstein lived in Princeton he kept up a continual correspondence with Queen
Elizabeth of Belgium. In these letters—some of the most
beautiful of all of Einstein's writings—one finds traced
most of the social and personal concerns of Einstein's
later years. The letters begin a month after Einstein's
arrival in Princeton in 1933 and end a month before his
death in 1955. Here are some of them.

November 20, 1933:

Dear Queen:
I should have written you long since, and would
have, were you not the Queen. Yet, I am not quite
clear why this fact should be an obstacle. But such
questions lie more within the province of a psychologist. Most of us prefer to look outside rather than
inside ourselves; for in the latter case we see but a
dark hole, which means: nothing at all.
Since I left Belgium I have been the recipient of
many kindnesses, both direct and indirect. Insofar
as possible I have taken to heart the wise counseling
of those who urged me to observe silence in political
and public affairs, not from fear for myself, but because I saw no opportunity for doing any good. . . .
Princeton is a wonderful little spot, a quaint and
ceremonious village of puny demigods on stilts. Yet,
by ignoring certain social conventions, I have been
able to create for myself an atmosphere conducive to
study and free from distraction. Here, the people who
compose what is called "society" enjoy even less
freedom than their counterparts in Europe. Yet, they

[3] Quoted in *Einstein on Peace*, p. 567.

seem unaware of this restriction since their way of
life tends to inhibit personality development from
childhood. Should civilization in Europe collapse as
it did in Greece, the intellectual desolation that will
result will be as profound as it was then. The tragic
irony is that the very quality which is the source of
the unique charm and value of European civilization—that of the self-assertion of the individual and
of the various nationality groups—may also lead to
discord and decay.[4]

On February 17, 1934, King Albert lost his life in a
mountaineering accident and was succeeded by his son
Leopold III. Einstein had heard that the now Queen
Mother was finding solace and consolation in her work
as an artist, and on February 16, 1935, he wrote:

. . . Among my European friends I am now called
"the Great Stone Face" [*der grosse Schweiger*: all of
Einstein's correspondence with Elizabeth—essentially all of his writing—was in German], a title I well
deserve for having been so completely silent. The
gloomy and evil events in Europe have paralyzed me
to such an extent that words of a personal nature do
not seem able to flow anymore from my pen. Thus I
have locked myself into quite hopeless scientific
problems—the more so since, as an elderly man, I
have remained estranged from the society here.

And, in reference to the queen's work in art, he goes on:

The effect of such work on us I know from my own
scientific endeavors. Tension and fatigue succeed
each other as they do if one is strenuously climbing a
mountain without being able to reach its peak. Intense preoccupation with things other than human
factors makes one independent of the vicissitudes of
fate; but it is a harsh discipline which reminds us
again and again of the inadequacy of our abilities.
At times I think back longingly to the happy hours
of the past and am tempted to visit Europe; but so

[4] *Ibid.*, p. 245.

many obligations would await me there that I seem unable to find the courage for such a project.[5]

By 1939 the fate of Europe and, above all, of his fellow Jews weighed so heavily on Einstein that he could no longer fully escape into the inner recesses of his work. Although a lifelong pacifist, he had come to the conclusion in the 1930s that armed resistance was the only way to confront Hitler and that the United States, despite its isolationist tradition, would inevitably be drawn into the war and should begin preparations for it. As he wrote to a peace meeting held in New York on April 5, 1938:

Many Americans, even pacifists, are thinking and saying: Let Europe fall, she deserves no better; we shall stand aside and have no part in it. I believe such an attitude is not only unworthy of Americans but shortsighted. It is unworthy of a great nation to stand idly by while small countries of great culture are being destroyed with a cynical contempt for justice. Such an attitude is shortsighted even from the point of view of enlightened self-interest. The triumph of barbarism and inhumanity can only lead to a situation in the world in which America herself will be forced to fight, and this under circumstances vastly more unfavorable than most people can possibly anticipate today.[6]

On January 9, 1939, Einstein wrote to Queen Elizabeth to ask her aid in obtaining permission for an aged cousin, in Germany, to emigrate to Belgium, and he adds:

I have been too troubled to write in good cheer. The moral decline we are compelled to witness and the suffering it engenders are so oppressive that one cannot ignore them even for a moment. No matter how deeply one immerses oneself in work, a haunted feeling of inescapable tragedy persists.

[5] *Ibid.*, p. 257.
[6] *Ibid.*, p. 279.

Still there are moments when one feels free from one's own identification with human limitations and inadequacies. At such moments, one imagines that one stands on some spot of a small planet, gazing in amazement at the cold yet profoundly moving beauty of the eternal, the unfathomable: life and death flow into one, and there is neither evolution nor destiny; only being.

The work has proved fruitful this past year. I have hit upon a hopeful trail, which I follow painfully but steadfastly in company with a few youthful fellow workers. Whether it will lead to truth or fallacy—this I may be unable to establish with any certainty in the brief time left to me. But I am grateful to destiny for having made my life into an exciting experience so that life has appeared meaningful.[7]

During the summer that followed Einstein went to Nassau Point, near Peconic, Long Island, where he sailed his boat and played chamber music with his neighbors. He wrote to Elizabeth on August 12, 1939:

Except for the newspapers and the countless letters, I would hardly be aware that I live in a time when human inadequacy and cruelty are achieving frightful proportions. Perhaps, someday, solitude will come to be properly recognized and appreciated as the teacher of personality. The Orientals have long known this. The individual who has experienced solitude will not easily become a victim of mass suggestion.[8]

And it was in his summer house on Long Island that Einstein signed his letter to Franklin Roosevelt warning of the implications of the discovery of nuclear fission. His letter to Roosevelt, written on August 2, just ten days before his letter to Elizabeth, was, in many ways, the formal initiation of the Nuclear Age.

[7] *Ibid.*, p. 282.
[8] *Ibid.*, p. 285.

No novelist could imagine a linkage of circumstances more unlikely than those leading from Einstein's birth on March 14, 1879, in the German city of Ulm to Nassau Point and the letter to Roosevelt, nor a personality more unlikely to have written it. In tracing out the steps that lead from Ulm to Long Island one is led to review the creation of nearly the whole of twentieth-century physics, since at nearly every step of the way there was the certain, guiding hand of Einstein.

Early Years

There was no precedent in Einstein's family history for any special scientific or intellectual achievement. As far back as he himself was able to trace the lineage on both his maternal and paternal sides, it consisted of typical merchants and artisans of German and, more generally, European Jewish life. Einstein's father, Hermann, was a rather happy-go-lucky, not very successful business man. When Einstein was a year old his father moved the family from Ulm to Munich, where he went into business with a brother with whom they shared a "double-house." The brother, Jakob, had some engineering training and took care of the technical side of the business: the manufacture of electrical equipment. From his mother, Pauline, Einstein acquired an early taste for classical music. She played the piano and he began taking violin lessons at the age of six

There were no visible signs of any special precocity in the very young Einstein. He was a dreamy child who disliked sports and games, talked with some difficulty (his parents were worried that he might be abnormal because it took him until he was three before he began to talk), and was dubbed *Pater Langweil*—"Father Bore"—by his nurse. A year after the family moved to Munich, Einstein's sister, Maja, was born. She was the only other child of the Einsteins.

Perhaps because of his solitary introspective childhood, Einstein appears to have retained a particularly vivid memory of it throughout his life. In a very real sense his feelings about the mysterious order that seems to underline the apparent chaos of natural events—the discovery that nature appears to present itself as a mathematical puzzle with remarkably simple and elegant solutions—were formed in his childhood. When he was in his early forties, in a conversation with the Berlin literary critic Alexander Moszkowski (they were discussing Newton's religious faith) he remarked:

> In every true searcher of Nature there is a kind of religious reverence; for he finds it impossible to imagine that he is the first to have thought out the exceedingly delicate threads that connect his perceptions. The aspect of knowledge which had not yet been laid bare gives the investigator a feeling akin to that experienced by a child who seeks to grasp the masterly way in which elders manipulate things.[1]

Although consistently agnostic with respect to any belief in a God preoccupied with the working out of human destiny, Einstein made throughout his life constant and amiable references to "God," whom he often called "the Old One." In this sense "God" stood for the rational connections, the laws, governing the behavior of the universe: both the fact that such laws seem to exist and that they are, at least to some degree, comprehensible by

[1] Alexander Moszkowski, *Conversations with Einstein*, p. 46.

us. In an essay written for a conference on science, philosophy, and religion in 1940 Einstein expressed his feelings as follows:

> The main source of the present-day conflicts between the spheres of religion and of science lies in the concept of a personal God. It is the aim of science to establish general rules which determine the reciprocal conceptions of objects in time and space. . . . It is mainly a program, and faith in the possibility of its accomplishment in principle is only founded on partial success. But hardly anyone could be found who would deny these partial successes and ascribe them to human self-deception. . . .
>
> The more a man is imbued with the ordered regularity of all events, the firmer becomes his conviction that there is no room left by the side of this ordered regularity for causes of a different nature. For him neither the rule of human nor the rule of divine will exists as an independent cause of natural events. To be sure the doctrine of a personal God interfering with natural events could never be refuted, in the real sense, by science, for this doctrine can always take refuge in those domains in which scientific knowledge has not yet been established. . . .
>
> To the sphere of religion belongs the faith that the regulations valid for the world of existence are rational, that it is comprehensible to reason. I cannot conceive of a genuine scientist without that profound faith. The situation may be expressed by an image: science without religion is lame, religion without science is blind.[2]

The two most vivid scientific impressions that Einstein preserved from his childhood were his discoveries of the behavior of the compass—the fact that by some mysterious attraction the compass needle always was made to point in a given direction—and, some time later, of the Pythagorean theorem of Euclidean geometry.

[2] Quoted in Philipp Frank, *Einstein: His Life and Times,* p. 285.

These two revelations are almost perfect illustrations of the complementary aspects of scientific phenomena. The behavior of the compass needle is a striking empirical fact which one might take, if so disposed, as evidence for the existence of magic. It requires a vast tradition of scientific experience even to imagine that such a phenomenon might have an explanation or, if one prefers a less loaded word than "explanation," a description in terms of generally applicable physical laws. On the other hand, the truths of Euclidean geometry appear as somehow self-evident, and it takes an equally vast tradition of scientific experience to discover how they fit in with the kind of empirical phenomenon illustrated by the compass. In any event, throughout his life Einstein made reference to the sense of "wonder" that the compass produced in him. When he was sixty-seven he attempted to explain this genesis of his scientific ideas:

> A wonder of such nature I experienced as a child of 4 or 5 years, when my father showed me a compass. That this needle behaved in such a determined way did not at all fit into the nature of events, which could find a place in the unconscious world of concepts (effect connected with direct "touch"). I can still remember—or at least I believe I can remember—that this experience made a deep and lasting impression upon me. Something deeply hidden had to be behind things. What man sees before him from infancy causes no reaction of this kind; he is not surprised over the falling of bodies, concerning wind and rain, nor concerning the moon or about the fact that the moon does not fall down, nor concerning the differences between living and non-living matter.[3]

[3] Paul Arthur Schilpp, ed., *Albert Einstein: Philosopher-Scientist*, p. 9. This volume in the Library of Living Philosophers is of collected essays by Einstein's old friends or colleagues of long standing. Einstein's introduction, where this passage appears, was often referred to by him as the writing of his own obituary.

Both in his elementary school and at the Luitpold *Gymnasium* in Munich Einstein was exposed to a certain amount of formal religious training. He was the only Jewish student in his Catholic elementary school and had received and, in fact, rather enjoyed the same instruction in Catholicism that the other students had He had no sense of anti-Semitism and no particulai attachment to Jewish rituals either. His family was entirely areligious, although they maintained their own version of the ancient Sabbath custom of inviting a poor Jew to share a meal with them. In the case of the Einsteins this took place on Thursday noon, when the family shared its meal with a poor Jewish student from Russia. This student, Max Talmey, introduced young Albert to some books on popular science which he read avidly.[4] At the same time in the *Gymnaisium* in Munich he began to receive, as was habitual at the time for Jewish students, special instruction in the Old Testament. For a while, somewhat to the amusement of his family, he developed almost fundamentalist views of the Bible. Very quickly these came into conflict with his studies in science. As he wrote in his "obituary":

> Through the reading of popular scientific books I soon reached [a] conviction that much in the stories of the Bible could not be true. The consequence was a positively fanatic [orgy of] freethinking coupled with the impression that youth is intentionally being deceived by the state through lies: it was a crushing impression. Suspicion against every kind of authority grew out of this experience, a sceptical attitude towards the convictions which were alive in any specific social environment—an attitude which has never again left me, even though later on, because of a better insight into the causal connections, it lost some of its original poignancy.[5]

[4] Talmey later wrote a book entitled *The Relativity Theory Simplified and the Formative Period of Its Inventor.*
[5] Schilpp, *op. cit.*, p. 9.

About this time his engineer uncle began giving Einstein informal lessons in algebra and geometry. Among other things, he told Einstein about the Pythagorean theorem, which, after considerable work, Einstein was able to prove. However, he had not realized the intricate logical framework of Euclidean geometry until he received a textbook from his uncle.

At the age of 12 I experienced a second wonder of a totally different nature: in a little book dealing with Euclidean plane geometry, which came into my hands at the beginning of a school year. Here were assertions, as for example, the intersections of the three altitudes of a triangle in one point, which— though by no means evident—could nevertheless be proved with such certainty that any doubt appeared to be out of the question. This lucidity and certainty made an indescribable impression upon me. That the axiom had to be accepted unproved did not disturb me. In any case it was quite sufficient for me if I could peg proofs upon propositions the validity of which did not seem to me to be dubious. For example I remember that an uncle told me the Pythagorean theorem before the holy geometry booklet had come into my hands. After much effort I succeeded in "proving" this theorem on the basis of the similarity of triangles; in doing so it seemed to me "evident" that the relations of the sides of the right-angled triangles would have to be completely determined by one of the acute angles. Only something which did not in similar fashion seem to be "evident" appeared to me to be in need of any proof at all. Also, the objects with which geometry deals seemed to be of no different type than the objects of sensory perception, "which can be seen and touched." This primitive idea, which probably also lies at the bottom of the well-known Kantian problematic concerning the possibility of "synthetic judgements *a priori,*" rests obviously upon the fact that the relation of geometrical concepts to objects of direct experience (rigid rod, finite interval, etc.) was unconsciously present.[6]

[6] *Ibid.,* p. 9.

In this Einstein alludes to the second element in the scientific experience mentioned earlier: the relation of what one may call "mathematical truths" to the "physical truths" disclosed by the examination of phenomena. We shall have occasion to come back to this in detail when we discuss Einstein's "general" theory of relativity with its geometrical interpretation of the phenomenon of "gravity." It is worth while here to make a comment about the significance of the word "paradox" in this connection, since it is often said that the theory of relativity is "paradoxical." In logic, or pure mathematics, a "paradox" is essentially equivalent to an inconsistency. It means that a given proposition does not follow, in a consistent fashion, from a set of axioms, which usually means that one has made a mistake in the process of logical deduction. It sometimes means that a conclusion one feels intuitively *ought* to follow from a set of axioms does not follow. On the other hand it is difficult to see what can be meant by a paradox in experimental physics. If the result of the experiment is correct, it is not a paradox but a *fact*. But some facts, newly discovered, seem to contradict one's intuitive sense of what the world should be like. Hence when people say that the theory of relativity is paradoxical what they usually mean is that it predicts things about the world which are unlike what they think the world, or nature, should be. The real question is not that the predictions match one's intuition, which is fallible, but that they be correct—that they lead to experimentally testable statements and that these statements pass experimental test. Einstein would, and did, argue that this description of the situation is something of an oversimplification. It ignores the fact that the general principles of physical theory are, in his words, "free creations of the human mind." This notion is very difficult to state with rigor— perhaps impossible. It has to do with the fact that the phenomena do not uniquely determine the theory. There may be several theories that give reasonable agreement with experiment. What the scientist hopes and, indeed, what he must assume is the case in order to motivate his work, is that there exists a theory, or theories, which because of their inner harmony and the compelling nature of their underlying assumptions bring one deeper into the workings of the universe—"closer to the Secrets of the Old One," in Einstein's phrase—than would a simple catalogue of the observed facts. Such theories may be formulated in terms of concepts that are connected to the observed facts only by long chains of deductive arguments, and because these concepts are to a certain degree "free creations of the human mind" we do, to some extent, impose the creations of our minds onto the workings of nature. We make experiments to fit our descriptions of nature as often as the other way around. These are very difficult matters to state with precision, and, traditionally, the views of scientists have ranged from that of the Austrian physicist Ernst Mach (who, as we shall see, played an important role in Einstein's intellectual development), who felt that theories were *simply* economical descriptions of observed fact, to the British astronomer Sir Arthur Eddington, who once wrote, "We have found a strange foot-print on the shores of the unknown. We have devised profound theories, one after another, to account for its origin. At last, we have succeeded in reconstructing the creation that made the foot-print. And Lo! it is our own."[7]

Apart from the work he did on his own Einstein hated the *Gymnasium*. He despised the rote learning and once remarked, "The teachers in the elementary school appeared to me like sergeants, and the *Gymnasium* teachers like lieutenants."[8] At age fifteen his life was sharply altered when his father's business in Munich failed and the family migrated to Pavia, near Milan, in Italy. Einstein was left behind to live with a family while

[7] Sir Arthur Eddington, *Space, Time and Gravitation*, p. 201.
[8] Cited in Frank, *op. cit.*, p. 11

attending the Luitpold *Gymnasium*, but after six months
he could no longer stand living alone in Munich and the
rigid discipline of school and, as Philipp Frank describes
it, manufactured a plan that would enable him to drop
out for a while. He got a physician to give him a
certificate stating that because of a nervous break-
down it was essential that he leave school and join his
parents in Italy. Then he obtained another certificate
from his mathematics teacher stating that his knowledge
of mathematics was sufficiently advanced so that he
could present himself for study at the university without the *Gymnasium* diploma.

> His departure from the gymnasium was ultimately
> much easier than he had anticipated. One day his
> teacher summoned him and told him that it would
> be desirable if he were to leave the school. Aston-
> ished at the turn of events, young Einstein asked
> what offense he was guilty of. The teacher replied
> "Your presence in the class destroys the respect [for
> me] of the students." Evidently Einstein's inner
> aversion to the constant drill had somehow mani-
> fested itself in his behaviour toward his teachers and
> fellow students.[9]

Liberated from the *Gymnasium*, Einstein joined his parents in Milan. One of his first acts was to renounce his
German citizenship. As he was a minor, this had to be
arranged by his father. Einstein had decided to become
a Swiss citizen, which was only possible at the age of
twenty-one. Hence he remained stateless from the age
of fifteen to twenty-one, which was, apparently, not
much of a problem in those days. Like many high-school
dropouts before and since, Einstein decided to spend
his time in part wandering around on foot. Unlike most
dropouts he also engaged in a program of mathematical
self-study. This was an extraordinarily happy period in

[9] Frank, *op. cit.*, p. 15.

his life and was only interrupted after a year when his father's business failed again and it became clear that Einstein would have to do something to support himself.

His father, apparently, felt that electrical engineering would be a suitable career. So it was decided that Einstein should go to Zurich, where, at the Swiss Federal Polytechnic School, there was the most celebrated center for the study of science in central Europe outside of Germany. This was something he had thought of doing even when he was in Munich, but in 1895 he failed the general entrance examination for the Polytechnic. Still, he had done so well on that part of the examination dealing with mathematics that the director of the Polytechnic suggested that he obtain a diploma at a Swiss cantonal high school and then reapply. Einstein enrolled in a school in Aarau which was run in a progressive fashion and gave the students an opportunity for independent work; above all, it had laboratories with good equipment which students could use to teach themselves science. After a year Einstein reapplied for entrance at the Polytechnic and, armed with a diploma, was accepted without further examination. It was in Zurich, when Einstein was sixteen, that he decided to abandon the study of pure mathematics which he had been teaching himself and to take up physics. In his "obituary" Einstein describes why he made this choice:

I saw that mathematics was split up into numerous specialties, each of which could easily absorb the short lifetime granted to us. Consequently I saw myself in the position of Buridan's ass which was unable to decide upon any specific bundle of hay. This was obviously due to the fact that my intuition was not strong enough in the field of mathematics in order to differentiate clearly the fundamentally important, that which is really basic, from the rest of the more or less dispensable erudition. Beyond this, however, my interest in the knowledge of nature was also unqualifiedly stronger; and it was not clear to

me as a student that the approach to a more profound knowledge of the basic principles of physics is
tied up with the most intricate mathematical methods. This dawned upon me only gradually after
years of independent scientific work. True enough,
physics also was divided into separate fields, each of
which was capable of devouring a short lifetime of
work without having satisfied the hunger for deeper
knowledge. The mass of insufficiently connected experimental data was overwhelming here also. In this
field, however, I soon learned to scent out that which
was able to lead to fundamentals and to turn aside
from everything else, from the multitude of things
which clutter up the mind and divert it from the
essential.[10]

It was also at this time, Einstein writes in his "obituary,"
that he began to realize that the foundations of the
physics he was learning, mostly by independent study
of the original texts, were fundamentally flawed. It took
the next ten years, until 1905, before he was able to sort
things out sufficiently in his own mind to be able to
write his first paper on the theory of relativity.

Classical Physics

In order to understand Einstein's insight we must review the state of physics as it presented itself to him at the end of the last century. In broad terms there were two main themes in physics: Newtonian mechanics on the one hand, and Maxwell's equations governing electricity and magnetism on the other.

The first science of mechanics, albeit incorrect, had its origins with the Greeks, and was systematized in the physics of Aristotle. According to this anthropomorphic physics earthly matter in motion sought the center of the earth, since this was its natural goal. (The motion of material objects, like those of conscious entities, was, according to Aristotelian physics, governed by motives and goals that varied according to the composition of the object; e.g., whether it was "earth" or "air.") Hence the more massive an object was the more rapidly it fell, since the greater was its tendency to seek the center of the earth. This, of course, reflects the common-sense observation that a rock falls, in air, more rapidly than a feather. The weakness of Greek physics was in the inadequate attention paid to quantitative experiment. (Some historians have attributed this to the social structure of Greek society, according to which such experiments would have been considered as menial activity of the sort usually assigned to servants or slaves.) For experiments would have revealed that this notion of the relation of the speed of fall of an object to its mass was false. A rock with a mass of ten grams does not fall, under similar circumstances, twice as rapidly as a rock with a mass of five grams. (The actual law of fall is complicated by the presence of the air, which causes the friction that affects the feather more than the rock.) In addition to the elements of earth, air, fire, and water which the Greeks believed composed the objects of common experience, they postulated a fifth essence—a "quintessence"—of which the heavenly bodies were alleged to be composed. These bodies, because of their special nature, were supposed to move only in perfectly circular orbits at constant speed –uniform circular motion.

This cosmology came into rapid conflict with the observations of planetary motion. The motion of the planets with respect to the "fixed" stars as observed from the earth is rather complicated, and nonuniform. In fact, a planet will even appear periodically to reverse its sense of direction in orbit—the phenomenon of "retrograde motion." To preserve the principle of uniform circular motion the later Greek astronomers were obliged to fit the observed orbits with very complicated superpositions of circular motions. (A modern scientist would call this data-fitting the expansion of the observed orbital motions in a Fourier series, after the early nineteenth-century French mathematician J. B.

Fourier, who showed that any regular periodic motion can be construed as a superposition of uniform circular motions, perhaps an infinite number of them.) This planetary orbital fitting was carried out in its greatest detail by the astronomer Ptolemy, who lived in Egypt in the second century A.D. His work was summarized in a book known as *Almagest* (*The Greatest*), which dominated European planetary astronomy through the Renaissance—until the time of the Polish cleric Copernicus early in the sixteenth century.

Copernicus discovered that if he changed the frame of reference of the planetary motions, so that the sun was at rest and the planets moved about it, he could simplify the description in terms of the superposition of circular motions. Copernicus was a firm believer in the Greek notion that the planetary motions had to be related to uniform circular motions, but he was able to reduce the complications (in modern language, to simplify the Fourier series) by changing the frame of reference to a sun-centered frame. However, it was not until the advent of Kepler a half century later that the principle of a description in terms of uniform circular motions was got rid of in astronomy once and for all. Kepler, making use of the precise data of his patron, the Danish astronomer Tycho Brahe, showed that the orbit of Mars could be fitted accurately by an ellipse with the sun at one focus. (One could decompose such an elliptical motion in a Fourier series, but this would be rather pointless since the elliptical orbit is so much simpler to visualize.) This was an enormous liberating step in the history of scientific ideas, but we must be clear, in order to understand the contribution of Newton, what it did *not* accomplish. Kepler's result was essentially an empirical observation. It had little predictive power. There was no way, for example, to account for why the planets moved in ellipse or why other objects, like projectiles, did *not*. (Kepler had the correct intuition that some influence emanating from the sun played a determining role in the characteristics of planetary orbits, but he never gave this idea a quantitative form.)

The next great figure in this story was Galileo. Of Galileo's many contributions to science we wish to focus on only two. The first was his conjecture that in the absence of air resistance all objects would fall to earth with the same acceleration independent of their mass. (We would now illustrate this by taking, say, a feather and penny, and putting them in a container from which we could evacuate the air, and showing that they do indeed fall at sensibly the same rate. Galileo did not have good vacua available to him so that he had to use somewhat more indirect methods. The legend that he dropped two cannon balls of greatly differing masses from the tower of Pisa to prove this is not, according to most historians, correct and, if carried out, would very likely have given the wrong answer.) The second contribution of Galileo upon which we wish to focus is a bit more subtle to describe. This has to do with the role of "inertia" in motion. According to Aristotelian physics the maintenance of an object in motion required the continuing action of what we would now call a force. This idea was abstracted from the common-sense experience that if we wish to push an object along the surface of the earth we must supply a force to keep the object moving. However, it is an equally common experience that once an object is set in motion it requires the action of a force to stop it or to change its direction. Here Galileo had the insight to imagine a situation in which all of the effects of friction could be removed—a very smooth surface like ice—which would have the property that once an object was set in motion, it would keep moving essentially indefinitely until a force was brought into play to alter its motion. Moreover, he understood that this "inertial" component of motion played a role in all of the motions that we commonly observe. As an illustration he considered the motion of a projectile in the absence of air friction. If the projectile is held at rest and dropped, it will fall straight down. On the other hand if the projectile is given an initial impulse in the horizontal direction, it will describe a curved trajectory —a parabola, in fact—before it returns to the ground. Galileo analyzed this motion as being compounded of two independently functioning components—a force that causes the projectile to drop in the vertical and an "inertial" component that would maintain the projectile on an infinite straight-line trajectory in the horizontal if the downward force could be switched off. In particular he understood that both a state of rest and a state of uniform motion in a straight line corresponded to a situation in which *no effective forces act*. In other words the role of the force is to *change* the state of motion or to produce what we would now call an *acceleration*. This was a very important observation and one that contradicted and replaced the erroneous Aristotelian analysis of forces and their relation to motion. The stage was now set for the Newtonian synthesis.

It is almost impossible to exaggerate the importance of Newton's influence in physics, and we will have a chance to discuss various aspects of it in more depth as we proceed. Here we would like to make clear how his work went beyond that of his predecessors, and why from the publication of the *Principia* in 1686 until the end of the nineteenth century Newton's theory of mechanics remained essentially unchallenged. To begin with, Newton made Galileo's analysis, which had been restricted to certain simple motions, both quantitative and general. He did this by changing the emphasis from a study of orbital motions as a whole to the study of "local" properties of orbits—the behavior of the motion from point to point along the orbit. Through his invention of the differential calculus (simultaneously and independently invented by Leibniz) he was able to give a precise definition of the rate of change of distance along an orbit when the intervals of length are allowed to become arbitrarily small. Hence he could define the velocity at any point along the orbit. Given the velocity he could define *its* rate of change at a point—i.e., the acceleration. It is this acceleration that is caused by whatever forces are acting on the object in orbit. Hence Newton arrived at what we would now call a "differential equation" which related force to acceleration—an equation that describes the motion over an infinitesimal portion of the orbit. (This is the equation $F = ma$ that every high-school student learns.[1]) In general, this equation is without content unless it is possible to specify the force.

The next step that Newton made was to produce a mathematical expression for the force of gravity. According to his notion every mass in the universe attracts every other mass with a force that is proportional to the product of the masses involved and which falls off as the square of the distance between them increases. It is now possible to insert this expression into the equation relating force to acceleration and then to solve this equation by the process of "integration," also developed by Newton. Integration enables one to sum up the effects of the infinitesimal pieces of the orbit. To visualize this, one may think of the orbit as being composed of tiny—infinitesimal—straight-line segments. The motion along these segments is easily computable, and the orbit as a whole can be reconstituted by summing up the line segments. The solutions to the equation yield the particle orbits, and Newton was able to show, with his expression for the gravitational force, that the only possible particle orbits for one particle moving under the gravitational influence of another—for example, a planet and the sun—are conic sections: ellipses, hyperbolas, and parabolas. Which of these orbits a

[1] Aristotle's equation of motion, expressed in modern form, would be
$$F = KV$$
where K is a constant and V the velocity. Hence, according to this force law, a body free of forces would, necessarily, be at rest.

particle selects depends on its "initial conditions"—
how much initial velocity is given to it. (All of this has
now become very familiar from rocketry, where we
know that to get a rocket in "orbit"—which means to
make it follow an elliptical path around the earth rather
than a parabolic path which will make it fall onto the
earth—we must supply the rocket with enough initial
velocity.) Hence with one stroke Newton recovered both
the elliptical orbits of Kepler and the parabolic projectile orbits of Galileo. However, the consequences of his
mechanics go much deeper. According to Newtonian
mechanics, once the forces and the initial conditions
are specified, it is possible to calculate the motions of
particles into the indefinite future. In other words, the
entire future course of the universe is both fixed and, at
least in principle, calculable, if we know its present
state and the forces. Later, when we discuss the quantum theory, we shall have occasion to re-examine this
notion from the point of view of modern physics. This
deterministic consequence of Newtonian mechanics was
succinctly stated by the Marquis Pierre Laplace, the
great French mathematician and physicist, whose five-
volume *Méchanique Céleste*, completed in 1825, summarized the development of mechanics following Newton's *Principia*. Laplace wrote:

> We must thus envisage the present state of the uni-
> verse as the effect of its previous state, and as the
> cause of that which will follow. An intelligence that
> could know, at a given instant, all the forces govern-
> ing the natural world, and the respective positions
> of the entities which compose it, if in addition it was
> great enough to analyze all this information, would
> be able to embrace in a single formula the move-
> ments of the largest bodies in the universe and those
> of the lightest atom: nothing would be uncertain for
> it, and the future, like the past, would be directly
> present to its observation.[2]

[2] *Essai philosophique sur les probabilités* (2nd ed 1014),
pp. 3–4.

Considering the uncanny success of Newtonian mechanics in the description of the motions of objects from planets to cannon balls, it is little wonder that there was a tendency for scientists to accept this theory, somewhat uncritically, as the last word and ultimate standard in scientific explanation.

We now turn to electricity. Although electricity in the form of spark discharges, lightning, and the like had been known as a somewhat eccentric curiosity for centuries, the first electrical studies in the modern scientific spirit began with the invention by Alessandro Volta, late in the eighteenth century, of the "voltaic cell" or battery. This device generates a steady current of electricity for long periods of time—until the energy source that is driving the current, a chemical reaction or whatever, is used up. Hence the properties of a steady flow of electricity could be studied. It was discovered unexpectedly by the Danish physicist Hans Christian Oersted, in 1820, that such an electric current would perturb a magnetic needle of the sort found in compasses. When the compass was placed near the electric current, the needle moved. Magnetism, the fact that pieces of some oxides of iron attract each other, had been known since the Greeks without anyone having an explanation for it, certainly without any notion that this apparently freakish phenomenon had anything whatever to do with electricity. The connection was made more quantitative when, soon after, the French physicist André Marie Ampère showed that a current flowing in a circular loop produced a magnetic force of exactly the same character as would be produced by an equivalent amount of magnetic material—say in an iron bar magnet. Hence he was led to make the conjecture, not widely accepted at the time, that the origin of magnetism in materials can be traced back to circulating electric currents in them. (This, we now know, is a very substantial portion of the truth, although there is an additional source of magnetism that has become apparent since the advent of the quantum theory, having to do with the intrinsic angular momentum—the "spin"—of the elementary charges.) He demonstrated that two current-carrying wires can have a magnetic interaction between them in the same fashion that two ordinary iron bar magnets produce a mutual force upon each other.

Hence by the beginning of the nineteenth century three things were clear with respect to electricity and magnetism:

1. Magnets influence, i.e., interact, with other magnets.
2. Electrical currents can interact with magnets.
3. Electrical currents can have magnetic interactions with each other.

The next great step was taken by the remarkable, largely self-taught English physicist Michael Faraday, whose engraved portrait Einstein kept on the wall of his study. Faraday, born in 1791, was a genius in experimental physics who came from a poor Yorkshire family —his father was a blacksmith—and received a minimal formal education, consisting, in his words "of little more than the rudiments of reading, writing and arithmetic at a common dayschool." At the age of twelve, Faraday went to work as an errand boy in a bookstore, and he began the independent study of the books on science that made their way into the shop. When he was nineteen the shop was visited by one Mr. Dance, who gave Faraday some tickets to attend the lectures in London of Sir Humphrey Davy, the great British chemist. Faraday was so impressed with Davy's lectures that he decided to work in science at no matter what menial level. He applied to Davy for some sort of job and as a proof of his earnestness submitted to Davy the notes, supplemented by diagrams he had made himself, of Davy's four lectures. He was engaged as a laboratory assistant in 1813 and began the series of experiments that led, eventually, to his discovery of electromagnetic induction and the concept of the electromagnetic field

We have already seen that by the early nineteenth century it was known that electrical currents produce magnetic forces. What Faraday showed, beginning in 1831, was that under certain circumstances magnets can produce electrical currents. In their most direct form Faraday's experiments consisted of the following: he took an ordinary iron bar magnet—a so-called permanent magnet, since it retains its magnetism more or less indefinitely provided that the bar is not heated or otherwise drastically tampered with—and moved it through a coil of conducting wire of the sort that was capable of carrying a current. (The doughnut-shaped coils that Faraday used are still extant, although from the photographs they look a bit worse for wear.) The coil was attached to a galvanometer that measured the flow of electrical current in it. When the magnet moved, the galvanometer registered a surge of current, indicating that a moving magnet could "induce" an electrical current. This unexpected discovery completed the symmetry between electricity and magnetism—although in a subtle form, since it is only a *changing* magnetic "field," as we would now say, that can produce a current. It is also the basic principle of the "dynamo," in which electric currents are produced by moving coils of wire—driven by coal or water power—through magnetic fields.

The notion of the electromagnetic "field" is also due to Faraday (he spoke of "lines of force"). This concept has its origin in Faraday's observation that if one takes bits of iron filings and puts them on, say, a sheet of paper so that they are movable, in the region of a bar magnet and then shakes or taps the paper, the filings will rearrange themselves so that they will form a pattern of lines that stretch from the north to the south pole of the magnet. Faraday abstracted from this observation the idea that these lines were, in effect, present even when the filings were not there. In other words the magnet produced a "field" of influence throughout space which could be measured at any point by observing the behavior of a bit of iron filing, or a small compass needle, at that point. He soon extended this notion to describe the influence of electrically charged objects on each other. Although this abstract idea of a "field" is an essentially mathematical one, Faraday did not have the mathematical training, or perhaps the mathematical skill, to make all of this into a quantitative theory. This was done, soon after, by the Scottish physicist James Clerk Maxwell. In a real sense one could say that Maxwell was to Faraday what Newton was to Galileo and Kepler.

Maxwell, born in Edinburgh in 1831, the year that Faraday discovered electromagnetic induction, was a mathematical prodigy who began doing serious original work in mathematics at the age of fourteen. When he was twenty-four, after a brilliant career at Cambridge, he was elected to the Chair of Physics at Marischal College in Aberdeen. His work covered essentially all branches of physics, and was of such quality that Einstein, throughout his life, expressed the feeling that Maxwell's contributions were more significant than his own. Early in his career Maxwell began the formulation of the set of equations that bear his name, which quantify Faraday's lines of force and are the starting point of all modern discussions of electricity and magnetism. These equations are not so simple to spell out for the nonspecialist as Newton's laws, since they are "partial differential" equations that require more background in calculus to follow than do the "ordinary differential" equations of Newton's law. The basic idea is, however, not difficult to state. As we have seen, a magnetic field that varies induces an electric-current flow. To describe this quantitatively we need an equation that relates the *variation* of a magnetic field to the induced current, or more generally, to *variations in the induced electric field*. A field can vary in both space and time; i.e., at a given spatial point the field can vary in time, or at a given time the field can have values that differ from point to point in space. The Maxwell equations relate the "partial variations" of, say, the electric field, in space, to the time variation of the magnetic field. They put into precise mathematical formulas just the sort of empirical observation that Faraday made. It is interesting that although Faraday did not have the mathematical training to follow the details of the Maxwell equations, he felt, instinctively, the power of the equations to express what he had been trying to say. When only twenty-six and still groping toward his final formulation, Maxwell received the following letter from Faraday, who was sixty-six:

> There is one thing I would be glad to ask you. When a mathematician engaged in investigating physical actions and results has arrived at his conclusions may they not be expressed in common language as fully, clearly, and definitely as in mathematical formulae? If so, would it not be a great boon to such as I to express them so?—translating them out of their hieroglyphics, that we also might work upon them by experiment. I think it must be so, because I have always found that you could convey to me a perfectly clear idea of your conclusions, which, though they may give me no full understanding of the steps of your process, give me the results neither above nor below the truth, and so clear in character that I can think and work from them. If this be possible, would it not be a good thing if mathematicians, working on these subjects, were to give us the results in this popular, useful, working state, as well as in that which is their own and proper to them?[3]

Faraday's reasonable request is one which contemporary physicists have sometimes tended to forget.

One of Maxwell's "conclusions" was the prediction of an entirely new phenomenon—the propagation of elec-

---

[3] Cited in, for example, D. K. C. MacDonald, *Faraday, Maxwell and Kelvin*, p. 79.

tromagnetic radiation *in the vacuum*. His idea was the following: if one can cause an electrically charged object to vibrate, then part of the electromagnetic field surrounding the charge will become detached and will propagate away from the charge as a wave. This wave, unlike sound waves or water waves, will, according to the Maxwell equations, propagate in empty space, i.e., in total vacuum. Moreover, from the equations, Maxwell could predict the speed at which these waves would propagate. He discovered that this speed was about 186,000 miles a second—the speed of light! This was the first clue that light was an electromagnetic phenomenon. We are so accustomed to the idea of light—and radio waves, another form of electromagnetic radiation —propagating through empty space from the stars, the moon, and all corners of the universe, that we hardly give any thought as to what a remarkable phenomenon it is when compared to the kinds of wave motion that are familiar to us in which we actually see some sort of material medium undulating. In fact Maxwell's contemporaries were inclined to disbelieve the whole affair, and the existence of such propagating electromagnetic waves in vacuum was only confirmed experimentally in 1888, nine years after Maxwell's early death, by the German physicist Heinrich Hertz, who invented oscillators to create the Maxwell waves and receivers to detect them. He was even able to confirm experimentally that the waves propagated with the speed of light. But physicists of this period, accustomed as they were to the mechanical models of Newtonian physics, would not accept the idea of a wave oscillating in nothing, and so they conceived a medium, the "ether," which was supposed to permeate all space and whose function was to provide some sort of substance in which the Maxwell waves could oscillate. As Maxwell himself put it, in 1865, "We have therefore some reason to believe, from the phenomena of light and heat, that there is an aethereal medium filling space and permeating bodies, capable of being set in motion and of transmitting that motion from one part to another, and of communicating that motion to gross matter so as to heat it and affect it in various ways."[4] As we shall see, the properties of this ether got stranger and stranger and much more difficult to visualize than the Maxwell waves themselves. Under the liberating influence of Einstein's ideas, the ether was dropped as a concept by most physicists fairly early in the present century.

We are now in a position to understand the significance of the puzzle that Einstein began struggling with at age sixteen, some seven years after Hertz's first experiment. He asked himself what would be the consequences of his being able to move with the speed of light. This question, innocent as it appears, eventually brought him into conflicts and contradictions of enormous depth within the foundations of physics. In the first place, according to Newtonian physics, a *Gedanken* experi-

[4] In James Clerk Maxwell, "A Dynamical Theory of the Electromagnetic Field," *Philosophical Transactions,* 155 (1865,), 459–512.

Maxwell did not *invent* the concept of an ether. In its modern form this can be traced back to Descartes. As a philosophical principle Descartes rejected "action-at-a-distance"—the idea that physical systems can interact with each other without some form of intermediary contact. Descartes held that this contact was propagated through a medium—the ether—and that light and heat, for example, were propagated by ether pressure. When the particle theory of light appeared to explain all the known phenomena, there was no need for an ether since action-at-a-distance was avoided by the intermediary propagation of these particles. However, after the particle theory of light was abandoned in the nineteenth century the problem of avoiding action-at-a-distance returned in an acute form and the ether theory was revived. In modern physical theories there is no action-at-a-distance, because, we believe, all of the known forces are propagated by the exchange of "quanta"—particles of a sort. Hence, the particle theory of light has been revived although in a new and vastly more sophisticated form.

ment—a thought experiment—involving a material observer actually moving, at least in principle, with the speed of light, is in no way prohibited. In fact, according to Newton's law, if one is accelerated long enough by the application of force, however small, one must eventually reach the speed of light and, indeed, any speed. But imagine a wave—for simplicity let us think of a wave with a regular pattern of crests and troughs. Suppose we are at rest and the wave moves by us. Then we will observe a regularly repeated pattern of crests and troughs. In other words, the amplitude of the wave motion as we observe it will go through periodic, repeated oscillations. In the case of the Maxwell waves these oscillations are just what is detected by the Hertz antennas. However, suppose we could move with the speed of the propagation of the wave. Then we could move alongside a crest, or a trough, and the oscillations would simply disappear to us, as moving observers. Now, according to the doctrine that had by then come into acceptance, light was just such an oscillatory wave motion in the ether. Hence if one could move with the speed of light in the ether, then to such an observer the wave pattern of light would apparently disappear. The Maxwell equations, however, do not provide for such a possibility, and hence either *they must be wrong or it must not be possible for a material observer to move with the speed of light.* From the point of view of classical physics either alternative seemed absurd.[5]

[5] It should be emphasized that both of the descriptions from which I have derived this account of Einstein's discovery of the theory of relativity—Philipp Frank's and Einstein's autobiographical notes—were written at least forty years after the fact. They make Einstein's discovery of the relativity theory comprehensible, in that once the problem is stated this way it becomes clear that the solution must involve some radical alteration of classical physics—either Newtonian mechanics or the electromagnetic theories of Maxwell and Hertz. What gives one pause is that there does not seem to be any written evidence that in fact this

Moreover, Einstein realized that an observer moving with the speed of light would be in a position to violate the "principle of relativity." To understand this we must re-examine an aspect of Newton's law discussed earlier but not emphasized. It will be recalled that in a situation in which no forces act on an object, such an object can be either at rest or in uniform motion in a straight line. A force is required only to produce an *acceleration*. Hence as far as the laws of mechanics are concerned there is no difference between a state of "rest" and a state of uniform motion in a straight line with a constant velocity. When one first confronts a statement like this it is tempting to dismiss it as obvious nonsense, whatever Newton's law says. In the motions with which we are familiar in daily life we generally begin from a state of rest and, by an application of some force, are accelerated—pushed or pulled—into a state of motion, where we "know" we are moving. But imagine the situation of being in a closed vehicle—a train—which has already been set into uniform motion; since there is no acceleration, nothing to shake us, we simply

---

was how Einstein viewed the problem *at that time.* On the contrary, Einstein wrote an essay when he was fifteen or sixteen (*"Uber die Untersuchung des Aetherzustandes im magnetischen felde,"* now published in the *Physikalische Blätter* 27 (1971), 385) in which he proposed experiments to test the theory of a mechanical ether. Even as late as 1901 he referred in letters to the "light ether" and to methods for testing the motion of matter with respect to it. There does not appear to be any document *from this period* that gives even a clue to the thought processes that led him to relativity, any document, apart from the 1905 paper, that shows the real transition from classical ether physics to relativity. Perhaps Einstein was considering both lines of thought simultaneously and, in his later reconstruction of the thought processes that led him to the correct formulation, left out the blind alleys that had not led him anywhere. I am grateful to Freeman Dyson for correspondence and discussion on this point, which because of its mystery only serves to emphasize the miracle of the final intellectual creation.

would have no way of detecting the movement. In fact the revolution of the earth around the sun gives us an approximate example. We are not conscious that it is happening because the accelerations due to the force of gravity are so small. The reason why this is called a principle of relativity is that it means, as far as Newton's law is concerned, there is no such thing as a state of absolute rest or absolute uniform motion. The only physically detectable—and therefore physically meaningful—states of uniform motion are the *relative* motions of one observer with respect to another. It is perfectly meaningful to say that one observer is moving uniformly with respect to another at a velocity of five miles an hour—we can, in principle, look out the window of the train and measure our velocity with respect to the ground—but it is not meaningful to say that the ground is absolutely at rest. At rest with respect to what? is the question.

Newton himself was aware of the difficulty of specifying states of absolute motion. When we discuss Einstein's general theory of relativity, we shall return to a more detailed discussion of Newton's analysis of this problem, without which the whole of Newtonian mechanics hardly makes sense. Here, we shall point out that Newton resolved the problem *theologically*. For him, a devout Christian mystic, it was enough that rest and motion were distinguishable in the consciousness of God. God, in other words, provides the absolute frame of reference in Newtonian mechanics. Because of the enormous practical success of Newton's theory, and because the "fixed" stars—which are not actually fixed but undergo motions that are very slight when viewed from the earth—provide a stationary frame of reference which is good enough for most practical problems in Newtonian mechanics, its theological underpinnings were largely forgotten or ignored by Newton's successors. However, the problem of relativity reasserted itself in a new guise with the discovery of Maxwell waves and the postulate of an ether to carry them. Could the ether, if it existed, provide the absolute frame of reference? As Einstein makes clear in articles, letters, and papers, it was obvious to him that such an idea was nonsense. In other words the Maxwell theory would have to satisfy a relativity principle as well. "From the very beginning it appeared to me intuitively clear that, judged from the standpoint of such an observer [one who is moving uniformly] everything would have to happen according to the same laws as for an observer who, relative to the earth, was at rest."[6] But here again the same paradox reasserted itself. If Newtonian mechanics were correct we could accelerate such an observer until he was moving at the velocity of light, and at this velocity light would no longer appear as light—i.e., as an oscillatory wave motion—in which case we could determine our absolute velocity in contradiction to the relativity principle. (Einstein had a particularly nice example to illustrate this. He imagined a man looking into a mirror illuminated by a light bulb. If the man and the mirror were to move with the speed of light, then according to the Newtonian physics the light from the bulb could never catch up with the mirror, so that at this velocity he would no longer be able to see himself. Hence he would be able to say that he was moving with the velocity of light in contradiction to the relativity principle.)

Before we discuss Einstein's resolution of this paradox —the special theory of relativity—it is important to make a brief historical detour. In an account of scientific history in terms of the life and ideas of a single man, especially so great a scientist as Einstein, it is easy to distort the perspective by giving the impression that he and he alone recognized and solved all of the basic problems. To avoid this it is essential to indicate what parallel and relevant developments in physics were tak-

[6] Paul Arthur Schilpp, ed., *Albert Einstein: Philosopher-Scientist*, p. 53.

ing place at the time Einstein was doing this work. I use the word "parallel" advisedly, since, from all evidence available, it is likely that Einstein was unaware of most of them. Einstein did not have a regular academic appointment in a university physics department until 1909, four years after his paper on relativity appeared, and while he was working on it, his nominal employment was as a minor official—a technical expert third class—in the patent office in Berne. Moreover, such was the isolation of the Polytechnic in Zurich, where Einstein received most of his formal training in physics, that no course was offered in which the Maxwell equations *were even taught*. He learned them himself by studying textbooks. It is also interesting to remark that in his paper on the special theory there is no specific reference to any other physics paper.

Now, in which way had these problems presented themselves to other physicists? As we have mentioned earlier, because of the prevailing mechanistic philosophy —the idea that the ultimate explanation of all physical phenomena lay in the construction of mechanical models for them—it was natural for the post-Maxwellian physicists to seek a mechanical model for the propagation of electromagnetic waves in empty space. In this picture an oscillating electric charge causes a disturbance in the "ether" and this disturbance propagates in a similar way to which sound propagates on the surface of a drum —by elastic vibrations of the material. However, it rapidly became clear, especially through the work of the Dutch physicist H. A. Lorentz, that the analogy was strained, to put it mildly. Sound waves and light waves differ in several important respects. For present purposes we shall focus on one. Light waves are "transverse": the plane in which the light wave oscillates is at right angles—perpendicular—to the direction of propagation of the wave. (Think of the wave oscillating up and down while it propagates forward.) Sound waves, which are formed by the compression and expansion of a material medium, can oscillate in the direction of propagation as well as transversely. In fact, what Lorentz showed[7] was that in order to have completely transverse light waves in the ether, the ether had to be *infinitely rigid*. Hence it had to be a medium which was both omnipresent, through which bodies could move without resistance, and at the same time infinitely rigid. As Einstein puts it in his "obituary,"

> The incorporation of wave-optics into the mechanical picture of the world was bound to arouse serious misgivings. If light was to be interpreted as undulatory motion in an elastic body (ether), this had to be a medium which permeates everything; because of the transversality of the lightwaves in the main similar to a solid body, yet incompressible, so that longitudinal waves did not exist. This ether had to lead a ghostly existence alongside the rest of matter, inasmuch as it seemed to offer no resistance whatever to the motion of "ponderable" bodies.[8]

But this was just part of the difficulty. A more remarkable "fact" about the ether emerged when the question was raised whether the earth was at rest or in motion with respect to it. By the end of the nineteenth century it had begun to appear absurdly as if neither alternative were possible.

The first hypothesis that was ruled out was that the earth dragged the ether along, like some sort of ghostly atmosphere, as it moved around the sun so that, in effect, the earth remained at rest within it. That this was not possible was decided rather early in the century by a consideration of the phenomenon of "stellar aberration," a result of the earth's orbital motion around the sun which was discovered by the British astronomer James Bradley in 1725. The textbook image that is useful in explaining aberration is the following: consider a man

[7] In a doctoral dissertation completed in 1875.
[8] Schilpp, *op. cit.*, p. 25.

walking with an umbrella in a rain shower and imagine that there is no wind so that the rain falls straight down. Since the man is walking, if he wants to avoid a rain drop in front of him he will have to tilt his umbrella at an angle to the rain. The faster he walks, with respect to the speed of the fall of the rain, the greater the angle through which he will have to tilt his umbrella. Now we can translate this image to the description of the "fall" of light into a telescope on the moving earth. In order to catch the light it is necessary to tilt the telescope through a small angle as compared to the "true" direction: i.e., the direction the star would have if the earth were at rest. Thus to see a star through a telescope attached to the moving earth one must look not directly at it, but slightly away from it. The amount of angular tilt in this case is very small, since the earth's speed in orbit is at most about one ten-thousandth of the speed of light.[9] The effect is observable because, as the earth is moving in an elliptical orbit around the sun, the telescope angle must be readjusted to take account of the constant change in its orientation. In fact, one must move the telescope through a tiny ellipse as the earth makes one complete revolution around the sun. However, if the earth dragged the ether along, this would act on the starlight like wind on falling rain and, in effect, would compensate for the earth's motion and cancel out the aberration effect. Since the aberration is observed and simply accounted for by the earth's motion one must conclude—and this *was* concluded—that the earth does not drag the ether.

This leaves open the second possibility, namely that the ether is at rest and the earth is moving through it

[9] This angle is given approximately by v/c, the ratio of the velocity of the earth to that of light. In this case it turns out to be about 20 seconds of arc. Recall that there are 360 "degrees" in a circle, and that each degree is split into 60 "minutes" and each minute into 60 "seconds." This was one of the early methods of measuring the speed of light.

and hence that the ether provides the absolute-rest frame needed for the Newtonian laws.[10] It turned out that this proposition could be subjected to a rigorous experimental test. These experiments were the work of another remarkable nineteenth-century figure, Albert Michelson, the first American to win a Nobel Prize in one of the sciences. (His was awarded in 1907, in physics, for his work in optics.) Michelson, who was born in Poland in 1852 and had been educated, after his family's migration to America, at the Naval Academy at Annapolis, began the various experiments on the properties of light that were to constitute his life's work in 1878. As was the custom, in fact the necessity, of the time, he traveled to Europe—in his case Germany and France—for advanced training in optics. With his new techniques he was able, in 1882, to give the most precise experimental value for the speed of light that had yet been obtained. (He found it to be 299,853 kilometers a second, or 186,320 miles a second. The techniques for measuring the speed of light have steadily improved; as of November 1972 the best value was given as 299,792.456.2 kilometers a second with a small error in the last place. In other words the speed of light is now known to a hundred-thousandth of a per cent!) By this time Michelson had returned to the United States and had joined the Case School of Applied Science in Cleveland as a physics professor. But it was in Germany that he had begun the development of his most important experimental instrument—the Michelson interferometer. The interferometer is a device that takes advantage of the wavelike characteristics of light. If two trains of waves are superposed then they will "interfere" with each other. This means that the two wave trains will combine to produce a resultant wave form whose characteristics are related to those of the original waves.

[10] There are intermediate possibilities in which the ether is only partially dragged along. See Max Born, *Einstein's Theory of Relativity*, for a good general survey.

In particular the original wave trains may or may not be "in phase." In other words, when two simple wave patterns of identical shape interfere, the two patterns may be exactly superposable on each other, or one of the waves may have its crests or troughs ahead or behind the other. In the latter case the resultant pattern will show characteristic "fringes." Now the Michelson version of the interferometer is, in principle, extremely simple. It consists of two straight "arms" that are joined at right angles. At the end of each arm there is a mirror. At the corner where the arms are joined there is a half-silvered mirror tilted in such a way that light from a source is "split"; i.e., half of it passes through the nonreflecting surface of the central mirror, while the other half is reflected at right angles along the second arm. The two light beams bounce back from their respective mirrors at the ends of the arms and meet again at the join, where they interfere with each other. If the time it takes for the light beams to travel back and forth on their respective arms is the same, then they will meet at the join in phase. If for some reason the two times are different, then when the beams meet they will be out of phase and hence will form observable "fringes." There are two reasons why the times could be different: either the two arms do not have the same length, so that it takes longer for the light to make the round trip along one arm than the other; or, even if they have the same length, the speed of light might be different in the different directions. This latter statement sounds rather odd, and from the point of view of a post-Einsteinian modern physicist it *is* odd. But from the point of view of a nineteenth-century physicist convinced of the existence of a stationary ether, one would, and *did*, reason as follows: as the earth moves through the ether—let us say, for simplicity, in a straight line with constant velocity—the ether appears to flow by, like water in a stream. In the stationary ether light moves with the speed discussed above. But to the observer on the earth, the ether is flowing by, and the light, once it begins propagating in the ether, will share its motion—like a swimmer caught in a current. In fact if the Michelson interferometer is placed so that one arm is in the direction of the current and the other at right angles to it the light moving along these two paths will propagate like a swimmer who swims back and forth, with and against a current, as compared to a swimmer who travels the same distance back and forth at right angles to the current. The effective light velocities will be different in these two cases, it was argued, and hence the time it takes to make the round trip along the two arms of the interferometer should be different, even if the two arms were *exactly the same length.*[11] Thus, if the earth were moving through the stationary ether one would expect to see interference fringes, and from these one could determine the absolute speed of the earth with respect to the ether.

The rather simple principles we have outlined here became the basis for one of the most celebrated experiments in the history of physics—the Michelson-Morley experiment. (Edward Morley was an American chemist and physicist who collaborated with Michelson on the first precise version of the experiment, although the idea was surely Michelson's, for he had carried out preliminary versions of it in 1881. What is usually called *the*

[11] The formulae for these two times were first worked out by Maxwell. If L is the length of each arm and v is the speed of the earth, it turns out that the time for the round trip along the arm perpendicular to the Earth's motion (c is always the speed of light in the ether) is given by:

$$t = \frac{2L}{c} \ \frac{1}{\sqrt{1 - v^2/c^2}}$$

while the time for the round trip along the arm parallel to the earth's motion is given by:

$$t' = \frac{2L}{c} \ \frac{1}{1 - v^2/c^2}$$

In general t is larger than t'.

Michelson-Morley experiment was performed in 1887.)
To get an idea of the precision involved we may remark
that if one carries out the straightforward algebra
required to find the difference of times of arrival of the
two light beams, under the conditions described above,
this turns out to be proportional to the *square* of the
velocity of the earth to that of light, which means one
part in *100 million*. Nonetheless, the experiment was so
well designed that it would have been able to detect
effects on the interference of this order of magnitude or
even less.[12] The fact was that *no such effects were
found*. To improve the reliability of the setup Michelson
and Morley had the interferometer mounted on a heavy
block of stone which was, in turn, mounted on disc of
wood floating in a tank of mercury. This enabled them
to rotate the whole apparatus constantly to eliminate
the possibility that they were seeing some accidental
effect caused by some quirk in the construction of one
of the arms.[13] They made the experiment in sixteen
different orientations of the arms. They made the experi-

[12] We emphasize here, again, that the Michelson-Morley
experiment measured an effect of order of magnitude $(v/c)^2$.
It was well understood at the time that all the terrestrial
measurements of the speed of light prior to Michelson which
were not accurate enough to exhibit effects of this order
of magnitude would not be affected by the earth's motion
with respect to the ether. Indeed, in 1879 Maxwell had
written, "In the terrestrial methods of determining the velocity of light the light comes back along the same path
again, so that the velocity of the earth with respect to the
ether would act on the time of double passage by a quantity
depending on the square of the ratio of the earth's velocity
to that of light, and this is quite too small to be observed."
Quoted in R. S. Shankland, "The Michelson-Morley Experiment," *American Journal of Physics*, 32 (1964), 17.
[13] Each arm was 11 meters long. They used yellow sodium
light with a wave length of 5.9×10^{-5} centimeters. From
Maxwell's theory, Michelson estimated that if they rotated
the arms through 90 degrees the interference pattern should
have shifted by about 4/10ths of the distance between two
fringes. No shift was observed.

ment at noon and at 6 p.m. to see if the earth's orientation to the sun had something to do with it. They also
planned to do the experiment every three months, to see
if some particular aspect of the earth's orbital motion
played a role.[14] But neither they, nor any subsequent experimenter—and it has been repeated with vastly greater
accuracy in the last decade, using modern electronic
techniques—ever found the slightest effect of the earth's
velocity with respect to the ether. (In the 1920s a physicist named Dayton Miller at the Mount Wilson observatory in Pasadena caused a brief stir when he seemed to
find a non-zero Michelson effect. This soon was dis-
proved when his experiment was shown to be wrong.)

[14] They wanted to eliminate the possibility that the whole
solar system might be in motion with respect to the ether
and that at some point along the earth's orbit the two motions might accidentally cancel out.

An Aside on the Michelson-Morley Experiment

iv

In most textbook presentations of the theory of
relativity one finds the statement that the
Michelson-Morley experiment was the foundation and starting point for Einstein's theory of
relativity. But there is no mention by name whatever of this experiment in Einstein's 1905 paper.
There is a vague reference to experiments of this
general character, but no reference specifically
to Michelson and Morley. One might be tempted
to conclude either that Einstein had never heard
of the experiment or that, if he had heard of it,
it made so little impression on him that he did
not bother to refer to it. Indeed, Einstein later
explained to various physicists that this was in
fact the case. Given the sort of man he was, it is
impossible to imagine that he was doing this to
call attention to his own ingenuity. Recently,

the physicist and historian of science Gerald Holton, who has had access to the Einstein archives—letters, notes, and documents—in Princeton, has been examining them for clues as to the relation of Einstein's work on the special theory and the Michelson-Morley experiment. He quotes from a letter that Einstein wrote to a historian in Illinois a year before his death:

Before Michelson's work it was already known that within the limits of the precision of the experiments there was no influence of the state of motion of the coordinate system on the phenomena, resp. their laws. H. A. Lorentz has shown that this can be understood on the basis of his formulation of Maxwell's theory for all cases where the second power of the velocity of the system could be neglected (effects of the first order).

According to the status of the theory it was, however, natural to expect that this independence would not hold for effects of second and higher orders. To have shown that such expected effect of the second order was *de facto* absent in one decisive case was Michelson's greatest merit. This work of Michelson, equally great through the bold and clear formulation of the problem as through the ingenious way by which he reached the very great required precision of measurement is his immortal contribution to scientific knowledge. This contribution was a new strong argument for the non-existence of "absolute motion," resp. the principle of special relativity which, since Newton was never doubted in Mechanics but *seemed* incompatible with electro-dynamics.

[Einstein goes on:] In my own development Michelson's result has not had a considerable influence. I even do not remember if I knew of it at all when I wrote my first paper on the subject (1905). The explanation is that I was, for general reasons, firmly convinced that there does not exist absolute motion and my problem was only how this could be reconciled with our knowledge of electro-dynamics. One can therefore understand why in my personal

struggle Michelson's experiment played no role or at
least no decisive role.[1]

By clarifying the role, or non-role, of the Michelson
experiment Einstein is calling attention to the fact that
in scientific creativity at this level there is not such a
simple connection between experiment and theory as is
sometimes assumed. "Intuition," the "free creativity" of
the mind, plays a decisive role as well. This is not to say
that the scientist as scientist is free to indulge in arbitrary fantasies about the universe. Ultimately all speculations, if they are meaningful, must result in experimentally testable propositions. But rather it is that
specific experiments themselves do not, in a simple way,
define the axiomatic basis for theory, and, in the creative
work of a great physicist, "intuition"—a feeling of how
the universe should be—plays a more important role in
formulating this axiomatic structure than the results of
any given experiment.

Whatever effect the Michelson-Morley experiment
may have had on Einstein, there is no question what
effect it had on his scientific contemporaries. They were
stunned. The whole mechanistic basis of the ether interpretation of the Maxwell equations was tottering. It was
a moment for heroic speculative efforts, and, in 1892, the
Irish physicist George Francis FitzGerald proposed a
remarkable explanation. His idea was that the arm of
the Michelson interferometer that was along the direction of motion of the earth *contracted* in just such a way
as to compensate for the time difference caused by the
different effective velocities of the light in the two arms.
According to FitzGerald's idea these two effects would
exactly cancel out and hence the null result of the
Michelson experiment would be explained. Of course, in
everyday life, we do not observe any such contraction of

[1] Gerald Holton, "Einstein and the 'Crucial' Experiment,"
969; see also Holton's "Einstein, Michelson and the 'Crucial
Experiment,'" p. 2.

objects in motion, but, in fairness to FitzGerald's idea, the contraction he proposed was an effect of order of magnitude of the square of the velocity of the moving object to that of light.[2] In fact to explain the Michelson experiment one would need a contraction of only 1/200 of a micron—about a hundred *millionth* of a meter—a length so small that only an interferometer measurement could possibly reveal it. The question was why should there be such a contraction other than as an *ad hoc* explanation of the Michelson experiment. In 1895 Lorentz, who had also hit on the notion of a contraction to explain the Michelson result, or non-result (as we shall see, such a contraction, with a totally different theoretical basis, is also a feature of Einstein's special theory of relativity, and it has come to be known as the Lorentz-FitzGerald contraction), proposed a tentative explanation for it. Lorentz had been engaged in developing a theory of electromagnetic forces to supplement Maxwell's theory of the fields. His basic idea was that electrically charged matter functioned as the source of the Maxwell fields, but that the fields existed in the empty spaces between the particles of matter. In other words he suggested a clear separation between matter and fields. In this picture two charged particles interacted with each other by means of the mutual influence of their respective fields. As Einstein put it in his "obituary":

> In principle a field exists, according to him [Lorentz], only in empty space. Matter—considered as atoms—is the only seat of electric charges; between the material particles there is empty space, the seat of the electromagnetic field, which is created by the position and velocity of the point charges which

[2] The exact formula for the Lorentz-Fitzgerald contraction is:
$$L = L_0 \sqrt{1 - v^2/c^2}.$$
Here L_0 is the length measured at rest, v is the velocity with respect to the ether, and c is the speed of light.

are located on the material particles. . . . The particle-charges create the field, which, on the other hand, exerts forces upon the charges of the particles, thus determining the motion of the latter according to Newton's laws of motion. . . . The physicist of the present generation regards the point of view achieved by Lorentz as the only possible one; at that time, however, it was a surprising and audacious step, without which the later development would not have been possible.[3]

The "Lorentz force" between electrically charged particles, which he derived from somewhat arcane considerations involving the ether, is still an essential ingredient in describing the interactions of charged particles like electrons. Lorentz had the idea that if matter consisted of "molecules," i.e., electrically charged bodies, held together by electromagnetic forces, it might well be that when such a body was set in motion these forces would change in such a way as to produce the Lorentz-Fitz-Gerald contraction. In 1906—a year *after* Einstein's paper had appeared—he gave a series of lectures at Columbia University in which he summarized this point of view.

We can understand the possibility of the assumed change of dimensions, if we keep in mind that the form of a solid body depends on the forces between its molecules, and that, in all probability, these forces are propagated by intervening ether in a way more or less resembling that in which electromagnetic actions are transmitted through this medium. From this point of view it is natural to suppose that, just like the electromagnetic forces, the molecular attractions and repulsions are somewhat modified by a translation imparted to the body, and this may very well result in a change of its dimension.[4]

[3] Paul Arthur Schilpp, ed., *Albert Einstein: Philosopher-Scientist*, p. 35.
[4] H. A. Lorentz, *The Theory of Electrons*, p. 201.

Physicists were still so much under the influence of the ether that Lorentz's ideas persuaded Morley and his colleague D. C. Miller (the same Miller who, some years later, thought he had found a non-zero Michelson effect) to undertake a new set of experiments in which they used first a wooden structure for the arms and then a steel structure, with the idea that if Lorentz's explanations were correct the effect could very well depend on which molecules the arms were made of. *The result was still zero.* The velocity of the earth with respect to the ether was unobservable.

Lorentz wrote up his lectures in book form in 1909, when he was fifty-six. (It was revised in 1915 with appendices and footnotes reflecting Lorentz's growing acceptance of Einstein's theory of relativity.) It is a classic document of nineteenth-century science. One can say "nineteenth-century science" because, although Lorentz was too great a physicist not to appreciate Einstein's work, it is clear that, at least in 1909, he could not quite get himself to believe it. At the end of the book Lorentz presents a summary of Einstein's theory, then just coming to the attention of his contemporaries, and after it he writes, "Yet, I think, something may also be claimed in favour of the form in which I have presented the theory. I cannot but regard the ether, which can be the seat of an electromagnetic field with its energy and its vibrations, as endowed with a certain degree of substantiality, however different it may be from all ordinary matter."[5] Lorentz's great book is something like a magnificent old European castle, beautifully constructed and yet somehow haunted by ghosts.

[5] *Ibid.*, p. 230.

Einstein and the Relativity of Time

V

While the Michelson-Morley experiment was occupying the attention of Einstein's scientific contemporaries, Einstein himself, as far as one can tell, remained nearly oblivious to it. Having failed to get any academic job of any sort in physics, he was earning his living in Berne examining patent applications for technical flaws; and, in such spare time as he had, was struggling with the real problem, which had (at least as he recalled it later) been apparent to him from the age of sixteen. From the point of view of Einstein's special theory of relativity the solution to the Michelson-Morley "puzzle" is so simple that, at first, one may be somewhat disappointed. It is—that there is no puzzle. The basic premise from which Einstein begins is that a state of rest and a state of motion with constant speed cannot be distinguished by any experiment, either electromagnetic or mechanical, performed by an observer in one system or the other. Of course the relative motions of the two systems can be measured. But each observer can claim, with equal validity, that he is at rest and it is the other observer who is moving. This is, in fact, just what the Michelson-Morley experiment confirms. The experiment is carried out on the earth, which to a high degree of approximation can be regarded as a system moving uniformly (such systems are known as "inertial systems"). Hence, no effect is to be expected. Of course, if Michelson had found an effect, the theory of relativity would simply have been wrong. In this context the Lorentz-FitzGerald contraction never enters the discussion, since there is no stationary ether to provide the absolute-rest frame. The real problem is how to reconcile Newtonian mechanics, which allows an observer to be accelerated to the speed of light, with the relativity principle for electromagnetic theory, which, as we have seen, cannot allow an observer to travel at the speed of light. The genius of Einstein was in recognizing that these two theories *cannot* be reconciled and that it is Newtonian mechanics that is wrong. Of course when one says that a theory like Newtonian mechanics—one of the most successful scientific discoveries ever made—is "wrong," what one means is that it is strictly correct only within a limited domain of phenomena. Newtonian mechanics was created to describe motions of objects that move much slower than the speed of light. For such objects Newtonian mechanics and the special theory of relativity give results that are nearly identical. There are, in principle, "relativistic corrections," but these are so small that in practice they can be neglected in, say, computations of planetary orbits. This explains why Newtonian mechanics works so well for so many astronomical phenomena, such as planetary motion, and why the relativity theory was not discovered sooner. It also explains why people have such difficulty with the theory. It appears, as we shall shortly see, to violate "common sense." But, of course, common-sense experience does not involve objects that move with speeds close to that of light. However, in cyclotrons, synchrotrons, and cosmic rays, for example, we deal with particles that move with speeds that differ by tiny fractions of a per cent from the speed of light. If one tried to treat the motion of such particles with classical Newtonian mechanics, one would get complete non-sense when one compared the results with experiment.

Einstein's first paper on relativity, which was pub-lished in the German journal *Annalen der Physik* in 1905 under the title "Zur Elektrodynamik bewegter Kör-per" ("On the Electrodynamics of Moving Bodies"), begins with an analysis of the relativity of time. In our ordinary experience we are conscious of an apparently irreversible flow of events with a division into what is occurring "now," what has occurred "before" and is con-tained in the memory, and what will occur in the "future." The fact that this subjective set of impressions is shared by the rest of the human race has led to the quantification of them—if one likes, to the "invention" of time. For purposes of physics we must be careful to dis-tinguish between the subjective sense of time, which, by its very character, must be imprecise and personal, and the "objective" time as measured by clocks. (No doubt the fact that individuals do not differ radically about their subjective sense of time is what inspired the intro-duction of clocks.) For our purpose a "clock" is any phe-nomenon that repeats itself: for example the periodic motion of a pendulum or a balance wheel, or even a heart beat. The more precisely it repeats itself, the more accurate is the clock. The first important observation of Einstein's 1905 paper is that every statement about the "objective" time of an event is, in reality, a statement about the simultaneous occurrence of two events; namely, the simultaneous occurrence of the event in question and, say, the superposition of the "hands" of a clock on the numbers painted on a dial. As Einstein put it, "When I say, for example, 'The train arrives here at 7,' that means that, 'the passage of the little hand of my watch at the place marked 7 and the arrival of the train are simultaneous events.'"[1] (The Polish physicist Leopold Infeld, who was one of Einstein's assistants in the 1930s, writes that this is "the simplest sentence I have ever encountered in a scientific paper."[2]) However, this sentence, as Einstein was quick to point out, contains an undefined concept, namely the concept of "simultaneity." Of course we all "know" what it means for two events to be simultaneous. Typically we look at the events and at our watches and compare the observations. In everyday life there is no need to analyze the procedure further. As a practical matter we function in this way without any special difficulties. However, if we think about it, we realize that because the speed of light is not infinite it takes a certain amount of time for the light that is illuminating the event in question to reach our eyes, and hence, strictly speaking, we are comparing an event that has already occurred with our watches. As a rule we can ignore this because light travels so fast and the distances involved are so small that this "delay" is irrelevant. However, if we want to time events on the moon with clocks on the earth, say, the delay is very significant. It takes light, or radio waves, which move at the same speed, about 2.5 seconds to make the lunar round trip. This raises the essential question of how we can tell if events are "simultaneous" on the earth and, say, the moon. We might be tempted to arrange matters as follows: first we synchronize our watches by comparing them when they have been brought arbitrarily close together to eliminate the delay effect. Now if the two watches have been constructed

---

[1] All of the English translations of Einstein's papers, unless otherwise indicated, are taken from *The Principle of Relativity*. For this passage from Einstein's 1905 paper, see p. 39.
[2] Leopold Infeld, *Albert Einstein, His Work and Influence on Our World*, p. 27.

identically, when we transport one to the moon we can, perhaps, have confidence that the watches have remained synchronous. But how can we really be sure? What is needed here is a procedure—an "operational definition," in the language of the physicist and philosopher of science Percy Bridgeman—of simultaneity.

In making a procedural definition of simultaneity Einstein took advantage of the fact that light propagation in the vacuum obeys a very simple law—namely, it propagates with constant speed in straight lines. Hence if we wish to check whether two clocks are synchronous we can proceed as follows: we can measure the distance between the clocks by, say, laying off rulers one on top of the other, and then we can, in principle, go to the mid-point between the two clocks and instruct observers stationed by these clocks to emit light signals when the respective clocks read, say, 7. If these light signals arrive at the mid-point "simultaneously," the clocks are synchronous at the instant of measurement. (In the relativity theory it is assumed that we can decide if two events that occur at the same space point are simultaneous.) This way we can construct a network of synchronous clocks. (A few years after the theory was published Einstein, who still did not have a regular academic job and was struggling to make a living, remarked, "In my relativity theory I set up a clock at every point in space, but in reality I find it difficult to provide even one clock in my room."[3]) All of this is elementary and the most conservative Newtonian physicist would find nothing disturbing about it. It is at the next step that the revolution begins. The question is, can this procedure be applied to two clocks that are in motion with respect to each other, and, if so, what is the result? Here Einstein made an hypothesis which at first sight is rather startling but which is confirmed by every experimental test ever made, namely, *the speed of light as measured*

[3] Philipp Frank, *Einstein: His Life and Times*, p. 76.

by an observer is the same no matter what speed the object emitting the light has, with respect to that observer, provided only that the light source is moving uniformly. This principle plays such an important role in Einstein's theory that it is worth while to state it as graphically as possible. What it says is that if we have a flashlight that emits light and we measure the speed of the light it emits, then this speed will always be the same no matter how rapidly the flashlight is moving with respect to us when it emits the light. (It is well-known that light emitted by a moving source is shifted in color— the so-called Doppler shift—which means that if the source is moving toward the observer the light is shifted toward the blue, which is to say that its frequency is increased and its wave length decreased, but these two shifts compensate in such a way that the speed remains the same.)

Direct experimental evidence can now be given for this "principle of constancy"; the most dramatic example is found in starlight coming from "double stars," pairs of stars that move in orbit around each other. Evidently there are places in the orbital motion where the star is moving toward the earth and places where it is moving away. If the speed of light differed at these two places in the orbit, one could easily prove that one would see various kinds of "ghost" images of the moving star. It would appear to be, so to speak, in two places at once. Such images are not observed, and this is conclusive evidence of the principle. What is remarkable about this example, and the others usually given, is that they came *after* the relativity theory. The double-star work, for example, was carried out by the Dutch astronomer Willem de Sitter in 1913. In formulating this principle Einstein was again guided by his "intuition" as to what was simple and correct. The Maxwell equations contain the principle of constancy as a feature, but, as they were in conflict with Newtonian mechanics, one had to "guess" as to which equations were right. In the early

1950s Einstein commented on this to the physicist R. S. Shankland, who had asked him about the principle of constancy. Shankland knew that there had been other suggestions at the time which experiments did not then rule out. These other suggestions, Shankland writes, Einstein

> gave up . . . because he could think of no form of differential equation which could have solutions representing waves whose velocity depended on the motion of the source. In this case the emission theory would lead to phase relations such that the propagated light would be all badly "mixed up" and might even "back up on itself." He asked me, "Do you understand that?" I said no and he carefully repeated it all. When he came again to the "mixed up" part he waved his hands before his face and laughed, an open hearty laugh at the idea.
>
> Then he continued, "The theoretical possibilities in a given case are relatively few and relatively simple, and among them the choice can often be made by quite general arguments. Considering these tells us what is possible but does not tell us what reality is."[4]

"Reality" comes from experiment but is interpreted by the "free creation" of the mind.

Armed with the principle of constancy, we can now return to the question of synchronizing a rest clock with a moving clock. As we have seen, this question comes down to whether two events that are seen as simultaneous by observers at rest will also appear simultaneous to moving observers. Let us imagine that we have a rest observer stationed at the mid-point between two clocks and that when we begin the experiment we have a moving observer just passing this mid-point. Both rest clocks read 7, and according to plan a light signal is fired from each clock. From the point of view of the mid-

[4] R. S. Shankland, "Conversations with Albert Einstein," p. 41.

point rest observer the two signals meet, simultaneously, a short while later. However, from the point of view of the moving observer *they do not*! Since he is moving toward one light signal and away from the other, the distance that the first signal travels before it meets him is less than the distance the second light signal travels, and since, according to the principle of constancy, both signals are moving with the same speed, the first signal will reach him before the second one. The moving observer will claim that the events are not simultaneous and that the clocks are not synchronized. Hence the moving observer and the rest observer will not agree on the settings of the clocks. "Time" in the moving frame will differ from "time" in the rest frame. There is nothing to be done about this. It is a consequence of the fact that light does not move with infinite speed. Until Einstein made this analysis it had been tacitly assumed that the rate of a given clock was the same whether or not the clock was in motion. Having raised this question, Einstein then went on, in his paper, to derive the mathematical formula that predicts how the rate of a given moving clock is related to the rest clock. This derivation uses both the principle of relativity and constancy. It is also extraordinarily simple from a mathematical point of view. The logic is subtle, but there is not a single mathematical formula that cannot be followed by someone with a knowledge of high-school algebra.

We shall take up further consequences of the special theory of relativity in the sequel, but here we make a few additional remarks about the relativity of time. First, we give a qualitative argument to show that the rate of moving clocks is *slower* than rest clocks. To this end, imagine a particularly simple form of clock. Consider two mirrors separated by a certain distance and imagine we have set off a light signal between the mirrors. The light will now bounce back and forth between the mirrors at a regular rate, since the speed of light is constant. (We can always imagine that the mirrors are in vacuum.) In principle this is a perfectly fine clock and we can make it as accurate as we like by decreasing the distance between the mirrors. Now suppose we attach the mirrors to the walls of something that can move so that the mirrors are in the vertical direction but we move the whole system in the horizontal at right angles to the line between the mirrors. Now we view this somewhat bizarre apparatus from the rest frame. Suppose the light starts off from the lower mirror. If the apparatus were at rest, then to hit the upper mirror the light would simply have to follow a straight-line path at right angles to the lower mirror. However, when the mirrors are in motion with respect to us, we will observe the light start off at an angle from the lower mirror to catch the upper one, which is moving. In fact, to make the round trip, as we view it from the rest frame, the light will have to follow a triangular path which is, evidently, *longer* than the path if the system is at rest with respect to us. Since according to the principle of constancy the speed of light is the same in both frames, we would say that the time of the round trip is longer for the "clock" that is in motion. Hence we would argue that the period of the light clock is longer when the clock is moving than when it is at rest. It is easy, using nothing more complicated than the Pythagorean theorem, to make this argument quantitative and to derive the mathematical expression relating the two clocks.[5] This expression is, and it must be, identical to

[5] It is straightforward to see that if the distance between the two mirrors is L the period of the rest "clock" is $2L/c$ while the period of the same "clock" moving with speed v is given by:

$$\Delta t = \frac{2L}{c} \ \frac{1}{\sqrt{1 - v^2/c^2}}$$

which is always longer than $2L/c$ and in fact becomes infinitely large as v approaches the speed of light c.

Einstein's general expression for the behavior of *any* clocks, since nothing in Einstein's arguments depends on how the clocks are constructed. One feature of this expression is especially interesting: namely, it ceases to make sense if the moving clock is moving faster than the speed of light. More exactly, the period of the moving clock becomes longer and longer, as seen by an observer at rest, as the clock's speed approaches that of light; and if the clock could attain the speed of light the period as observed by the rest observer would become *infinite*. Hence the speed of light is a natural speed limit in the theory, and this is what is needed to avoid the paradox that Einstein discovered at the age of sixteen.

We should emphasize here that according to the relativity principle, an observer stationed *on* a uniformly moving clock will not notice any effects of his motion. Hence, every such observer will claim that *his* clock is measuring the "true" or "proper" time and that it is the clock in the frame moving with respect to his own that is slow.

While these arguments are perfectly correct, one should not rush out to test them by running past someone's wrist watch. As we have seen, these relativistic effects are, typically, of an order of magnitude of the square of the speed of, in this case, the clock to the speed of light. Using Einstein's formula one can show that, for example, if the clock moved at *half* the speed of light, i.e., 93,000 miles a second, it would appear to be slowed down by about 13 per cent. Hence for ordinary terrestrial speeds of even a few hundred miles an hour the effect is from a practical point of view negligible.

In his paper Einstein gave a rather fanciful example to emphasize the principle. He imagined two identical clocks, one at the North Pole and one on the equator. He then pointed out that the clock on the equator would have a rate a tiny fraction of a per cent slower, as measured by the clock on the pole, because of the earth's rotation. However, lest one get the impression that all of this is idle speculation, it is important to realize that in the high-energy physics of elementary particles this time retardation plays a crucial and readily observable role. Most of the so-called elementary particles are unstable. They decay into stable particles. These decaying particles have what is called a "half life"—the amount of time it takes half of any given sample to decay away. This half life is also the period of a kind of clock, and hence, according to the theory, a sample that is in motion should decay with a longer half life than a sample at rest. In the big accelerators one makes such particles by the millions, and when these emerge from the accelerator they are often moving at speeds that differ from the speed of light by tiny fractions of a per cent. It is possible to measure the half life of such a beam and to compare it with the half life of an equivalent sample of the same particles that have been brought to rest and then decay. In this case the two "half lives" differ very substantially, and the result is in perfect agreement with the theory of relativity. (For convenience, physicists call *the* half life the half life of the sample at rest.)

No doubt a reader coming upon these ideas for the first time must feel a certain sense of bewilderment at the apparent complexity and subtlety of the natural universe. And, one would hope, he will also feel the sense of beauty and even "simplicity" in the laws that underlie its behavior. As Einstein himself expressed it, *"Raffiniert ist der Herr Gott, aber boshaft ist er nicht."* ("God is sophisticated, but not malicious.") Einstein, throughout his life, never tired of trying to simplify and make more beautiful the formulation of his ideas. In this respect the following anecdote is typical.[6] In 1943 Einstein was approached by the Book and Author Committee of the Fourth War Loan drive with the suggestion that, if he still had the original manuscript of his 1905

[6] I am grateful to Helen Dukas for telling me of this incident.

paper on the special theory of relativity, and if he did
not object, it might be auctioned off at a war-bond rally
to raise money for the war effort and then placed in the
Library of Congress. Einstein was in principle perfectly
agreeable, but the original manuscript had been thrown
away in Berne in 1905 just after the paper was written,
since at the time he did not have any notion that it
would be of value. However, he did have the manuscript of a new paper, "Bivector Fields II," written in
collaboration with Professor V. Bargmann of Princeton.
(On February 4, 1944, this manuscript was sold at a war-
bond auction in Kansas City for five million dollars to
W. T. Kemper, Jr., a custodian of insurance funds, and
then placed in the Library of Congress.) He presented
the new manuscript and then, sensing the disappointment of his visitor, had an idea. He would copy the
entire thirty-page relativity article from the 1905
Annalen der Physik by hand and it could be sold as a
facsimile. At the same rally it was sold to the Kansas
City Insurance Company for six and a half million dollars and turned over to the Library of Congress.) He
asked Helen Dukas, his secretary, to read the manuscript to him, since he had forgotten exactly how he had
put the various arguments. Einstein looked up at one
point as she was reading to ask if that was actually
what he had written. Assured that it was, he remarked
that he realized now that he could have said it more
simply.

2. Relativity, Gravitation, and Cosmology

Preamble: Einstein When Young

vi

On November 28, 1915, the German physicist
Arnold Sommerfeld in Munich received a response to several unanswered letters that he had
written to Albert Einstein, who was then in
Berlin. Einstein wrote:

> During the last month I experienced one of
> the most exciting and most exacting times of
> my life, true enough, also one of the most suc-
> cessful. [Letter-] Writing was out of the ques-
> tion.
> For I realised that all of my field equations
> of gravitation up till now were entirely without
> support. Instead of that the following points
> of departure turned up. . . .
> After all confidence in the former theory
> has thus disappeared I saw clearly that a
> satisfactory solution could be found only by
> means of a connection with the universal
> theory of covariants of Riemann. . . . The final
> result is as follows. . . .

Now the marvelous thing which I experienced was the fact that not only did Newton's theory result as first approximation but also the perihelion motion of Mercury (43″ per century) as second approximation. For the deflection of light by the sun twice the former amount resulted.[1]

Einstein had just succeeded in creating the "general theory of relativity," as opposed to the "special theory of relativity," the subject of his 1905 paper. This theory, which many physicists believe to be the most perfect and aesthetically beautiful creation in the history of physics, perhaps in all of science, has replaced Newton's theory of universal gravitation. It has cleared up some anomalies in the behavior of planetary orbits—the "perihelion" of Mercury; led to significant predictions— the fact that light rays are bent by the gravitational attraction of the sun; become the basis for all of modern cosmology, including the expanding universe; and is at the present moment—with the discovery of pulsars and the conjectured existence of "black holes" of gravitation —again at the center of scientific interest. Its methodology is so different from anything that has gone before or since that even now, over a half century later, it is not clear how to fit it in with the rest of physics. It is small wonder that, in Sommerfeld's words, he reacted "somewhat incredulously." To which Einstein replied on February 8, 1916, on a postcard, "Of the general theory of relativity you will be convinced, once you have studied it. Therefore I am not going to defend it with a single word."[2]

At the time of this correspondence Einstein was thirty-six. He was by then widely recognized by his scientific contemporaries as a creative genius of the first rank. Indeed, a few years earlier Max Planck, the progenitor of the quantum theory and one of the leading

[1] Paul Arthur Schilpp, ed., *Albert Einstein: Philosopher-Scientist*, p. 100.
[2] *Ibid.*

physicists in the world, had written a letter of recommendation for Einstein, who was then being considered
for a job in Berlin, containing the sentence, "If Einstein's
theory [the "special theory of relativity"] should prove
to be correct, as I expect it will, he will be considered
the Copernicus of the twentieth century," and, in fact,
the public at large was beginning to recognize and take
interest in his work.[3] Einstein's personal fortunes had
also improved considerably from his patent-office days
in Berne, where he had done his physics in his spare
time after an eight-hour working day. He was now the
director of the newly formed Kaiser Wilhelm research
institute for physics and a member of the Royal Prussian
Academy of Science, with the title of professor at the
University of Berlin. (He had no official academic obligations and could divide his time between research and
teaching in any way he saw fit.) All of this had taken
place in 1913. From all accounts none of it had the
slightest effect on Einstein's basic way of life. He had
no interest in the social events which were meant to be
an essential part of the life of a great German
Geheimrat, and, in every way, according to Philipp
Frank, who saw a fair amount of him, he resembled no
one so much as one of those Bohemian violinists who
frequented the cafés and coffee houses where both he
and Professor Frank spent a good deal of their spare
time. In his book Frank recounts the following anecdote,
which is, no doubt, typical:

> Einstein was always very intent on being a person
> who did not require any special consideration. On
> one occasion he was supposed to pay his respects to a
> member of the Berlin Academy. He was not very
> fond of such formal visits, but he had heard that
> Professor Stumpf, a well-known psychologist, was
> greatly interested in the problem of space perception.
> Einstein thought that he would be able to discuss
> matters of mutual interest that might have some

[3] See Philipp Frank, *Einstein: His Life and Times*, p. 101.

connection with the theory of relativity, and he decided to make this call. On the chance that he might
find the professor at home, he went there at eleven in
the morning. When he arrived the maid told him
that the *Herr Geheimrat* was not at home. She
asked Einstein whether he would like to leave a mes-
sage, but he said it was unnecessary. He did not
want to disturb anyone and would come back later
in the day. "In the meantime," he said, "I'm going
to take a walk in the park." At two in the afternoon
Einstein returned. "Oh," said the maid, "since you
were here the *Herr Geheimrat* came home, had his
lunch, and because I did not say that you would
come back, he is taking his afternoon nap." "Never
mind," said Einstein, "I'll come back later." He
went for another walk and came back again at four.
This time he was finally able to see the *Geheimrat*.
"You see," Einstein said to the maid, "in the end
patience and perseverance are always rewarded."
. . . The *Geheimrat* and his wife were happy to see
the famous Einstein and assumed that he was now
making his formal introductory visit. Einstein, however, immediately began to talk about his new
generalization of the relativity theory and explained
in detail its relation to the problems of space. Professor Stumpf, who was a psychologist without any
extensive mathematical knowledge, understood very
little of the discussion and was hardly able to put a
word in edgewise. After Einstein had talked for
about forty minutes, he remembered that he was
actually supposed to be paying an introductory call,
and that it had already lasted too long. Remarking
that it was quite late already, he departed. The professor and his wife were dumfounded for they had
had no opportunity to ask the customary questions:
"How do you like Berlin?" "How are your wife and
children?" and so on.[4]

I once asked Professor Frank whether the Einstein of
this period would have seemed "bright" if one met him
in conversation or in a physics colloquium. In asking this

[4] *Ibid.*, pp. 115–16.

apparently manic question, I had in mind that even the
greatest of physicists come in all varieties: some are incredibly quick—some are not; some take themselves extremely seriously—and some do not; some are full of
humor—and some are not. I was curious to know where
the young Einstein fitted into this spectrum. The Einstein of the postwar years in America—he was then in
his sixties—had already assumed, as naturally as he had
assumed everything else, the aura of an Old Testament
prophet. From his pictures he seemed to carry, in his
eyes, the burdens of the world, like one of the thirty-six
Just Men of the Talmudic legends into whose hearts
God has poured "as into one receptacle all our griefs."
But in his early days he was, as Professor Frank put it,
"very bright" and much given to jokes and "cracks."
There was an irrepressible sense of joy and gaiety about
him, combined with a profound sense of inner serenity
and an all but impenetrable wall of reserve.

The immediate impression that Einstein made on
his environment was a conflicting one. He behaved
the same way to everybody. The tone with which he
talked to the leading officials of the university was
the same as that with which he spoke to his grocer
or to the scrubwoman in the laboratory. As a result
of his great scientific discoveries, Einstein had already acquired a profound inner feeling of security.
The pressure that had often burdened his youth was
gone. He now saw himself in the midst of the work
to which he was going to devote his life and to which
he felt himself equal. Alongside this work the problems of daily life did not appear very important.
Actually he found it very difficult to take them seriously. His attitude in intercourse with other people,
consequently, was on the whole one of amusement.
He saw everyday matters in a somewhat comical
light, and something of this attitude manifested
itself in every word he spoke; his sense of humor was
readily apparent. When someone said something
funny, whether intentionally or not, Einstein's re-

sponse was very animated. The laughter that welled
up from the very depth of his being was one of his
characteristics that immediately attracted one's attention. To those about him his laughter was a
source of joy and added to their vitality. Yet sometimes one felt that it contained an element of criticism, which was unpleasant for some. Persons who
occupied an important social position frequently had
no desire to belong to a world whose ridiculousness
in comparison to the greater problems of nature was
reflected in this laughter. But people of lesser rank
were always pleased by Einstein's personality.

Einstein's conversation was often a combination
of inoffensive jokes and penetrating ridicule, so that
some people could not decide whether to laugh or to
feel hurt. Often the joke was that he presented complicated relationships as they might appear to an in-
telligent child. Such an attitude often appeared to be
an incisive criticism and sometimes even created the
impression of cynicism. Thus the impression Einstein made on his environment vacillated between
the two poles of childish cheerfulness and cynicism.
Between these two poles lay the impression of a very
entertaining and vital person whose company left
one feeling richer for the experience. A second gamut
of impression varied from that of a person who
sympathized deeply and passionately with the fate of
every stranger, to that of a person who, upon closer
contact, immediately withdrew into his shell.[5]

Despite what Professor Frank often referred to as
Einstein's "rabbinical" appearance, he retained at least
an aspect of this mischievous and puckish side of his
personality to the end of his life. Abraham Pais, a colleague of Einstein's in Princeton, visited Einstein regularly from 1947 till about half a year before his death.

We discussed physics; very often it concerned the
foundations of quantum mechanics. On a number
of occasions Gödel, the logician, would join. Even
though there was not much of an agreement, I would

[5] *Ibid.*, pp. 76–77.

always come away feeling better for these talks. On one occasion I told Einstein a joke to which he responded with one of the most extraordinary kinds of laughter I have ever heard, then or since. It was rather like the barking of a seal. It was a happy laughter. From that time on, I would save a good story for our next meeting, for the sheer pleasure of Einstein's laugh, which would light up his face and make him look almost like a boy enjoying a good prank.[6]

Let us return to Einstein's early life and to the steps that led to his appointment in the patent office in Berne. As I have mentioned, Einstein had been admitted to the Polytechnic in Zurich in 1896, where he had an erratic career mostly of self-education, much of it doing experiments in the laboratory. Unlike many, if not most, theoretical physicists, Einstein seemed to be quite at home in the laboratory and retained a fondness for scientific gadgets throughout his life. (Einstein never developed any interest in chess or mathematical puzzles—common forms of recreation for theoretical physicists—but he liked figuring out how new inventions worked, and one of his most delightful popular essays, written in 1925, deals with the workings of a newly invented sailboat—a strange craft called, after its inventor, the Flettner ship, in which the wind acts on vertical rotating sheet-metal cylinders, which function as sails.)

Late in his life, Einstein reminisced about his experience at the Polytechnic:

There I had excellent teachers (for example, Hurwitz Minkowski [who, ironically, was later to make significant contributions to the mathematics of the theory of relativity]), so that I really could have gotten a sound mathematical education. However, I worked most of the time in the physical laboratory, fascinated by the direct contact with experience. The balance of the time I used in the main in order to

[6] I am grateful to Professor Pais for this description.

study at home the works of Kirchoff, Helmholtz, Hertz, etc. . . . The hitch in this was, of course, the fact that one had to cram all this stuff [*"all diesen Wust in sich hineinstopfen musste"*] into one's mind for the examinations, whether one liked it or not. This coercion had such a deterring effect [upon me] that, after I had passed the final examination, I found the consideration of any scientific problems distasteful to me for an entire year. In justice I must add, moreover, that in Switzerland we had to suffer far less under such coercion, which smothers every truly scientific impulse, than is the case in many other localities. There were altogether only two examinations; aside from these, one could just do as one pleased. This was especially the case if one had a friend, as did I, who attended the lectures regularly and who worked over their content conscientiously. This gave one freedom in the choice of pursuits until a few months before the examination, a freedom which I enjoyed to a great extent and have gladly taken into the bargain the bad conscience connected with it as by far the lesser evil. . . . It is, in fact, nothing short of ` miracle that the modern methods of instruction have not yet entirely strangled the holy curiosity of inquiry; for this delicate little plant, aside from stimulation, stands mainly in need of freedom; without this it goes to wrack and ruin without fail. It is a very grave mistake to think that the enjoyment of seeing and searching can be promoted by means of coercion and a sense of duty. To the contrary, I believe that it would be possible to rob even a healthy beast of prey of its voraciousness, if it were possible, with the aid of a whip, to force the beast to devour continuously, even when not hungry, especially if the food handed out under such coercion were to be selected accordingly.[7]

Among Einstein's student friends was Mileva Maritsch (Maric), a woman of Serbian and Greek Orthodox background who had come from Hungary to study in Zurich and whom Einstein married in 1903.

[7] Schilpp, *op. cit.*, pp. 15–18.

They were both students in the section of the Polytechnic whose primary function was in the training of science schoolteachers. In other words Einstein was not
studying to be a physicist but, rather, a high-school
physics teacher. He was receiving 100 Swiss francs a
month ($25) from a relative, of which he put aside 20
toward paying for a Swiss citizenship which he was
finally able to obtain from the canton of Zurich in 1901.
His only hope of going on for advanced training was to
obtain some sort of assistanceship from one of the professors at the Polytechnic or elsewhere. This was frustrated when his teachers failed to recommend him for
such a position after his graduation in 1900. He managed to find a temporary teaching job in a technical
high school in Winterthur, near Zurich, but this lasted
only a few months. He next answered a "want ad"
placed in a newspaper by a teacher in Schaffhausen
who was looking for someone to tutor young students in
a boarding school that he ran. This job also had a rather
limited tenure, since Einstein managed to convince his
charges that the *Gymnasium* education that they were
receiving was stifling, and when he requested that he
be given the full responsibility for the education of the
boys he was fired. It was at this time, 1901, that Einstein applied for employment in the patent office. He
was taken on, on a probationary oasis, in June 1902
and remained for the next seven years.

He now had enough financial security to marry. Ultimately the marriage was not successful—the Einsteins
were separated in 1914 and finally divorced in 1919.
Accounts differ somewhat as to what kind of a person
Mileva was. Professor Frank, who must have seen a
good deal of her and who was a shrewd judge of character, writes:

> She was somewhat older than he. Despite her Greek
> Orthodox background she was a free-thinker and a
> progressive in her ideas, like most Serbian students.
> By nature she was reserved, and did not possess to

any great degree the ability to get into intimate and pleasant contact with her environment. Einstein's very different personality, as manifested in the naturalness of his bearing and the interesting character of his conversations, often made her uneasy. There was something blunt and stern about her character. For Einstein life with her was not always a source of peace and happiness. When he wanted to discuss with her his ideas, which came to him in great abundance, her response was so slight that he was often unable to decide whether or not she was interested.[8]

They had two sons, Hans Albert, born in 1904, and now a professor of hydraulic engineering at Berkeley, and Eduard, born in 1910, who died recently. In an interview published a few years ago Hans Albert Einstein took some exception to Professor Frank's characterization of his mother and replied, "Stern? Severe? This is not, I believe, really correct. A person who had gone through all kinds of mishaps and so on, but not really severe. I would say able to give . . . and in need of love. That means somebody not essentially based on intellect."[9]

The picture that emerges from these early years is of a largely self-educated young man, unbeholden to either a religious or a national tradition, and above all accustomed, almost from childhood, to making and acting on his own judgments. During his years in the patent office he worked under conditions which a contemporary scientist would find all but impossible. He had contact neither with professional physicists nor with the books and journals he needed for his work, since these were not available at the Patent Office or even at the library at the University of Berne. He had no guidance from senior colleagues and no encouragement either. In physics he had to be self-reliant. There was no one else to rely on.

[8] Frank, op. cit., p. 23.
[9] The New York Post, May 23, 1963, p. 27.

Lorentz and Poincaré

vii

It is doubtful that a contemporary physicist could get a paper published that was written in the style of Einstein's 1905 work on relativity. Despite the fact that nearly every idea, and many of the formulas, bear at least a distant familial relationship to work that had been done by others—most notably Lorentz and Henri Poincaré—there is not a single reference to any of them. In fact the similarity—more apparent than real—between Einstein's work and those of his immediate predecessors, which would be caught by any competent referee of a good physics journal, has even led some historians of science to denigrate Einstein's contributions completely. The most remarkable example is Sir Edmund Whittaker, a distinguished British mathematical physicist who wrote a two-volume study on the history of the theories of ether and electricity in which Einstein's contribution to the theory of relativity is summarized by the sentence, "In the autumn of the same year [1905] . . . Einstein published a paper which set forth the relativity theory of Poincaré and Lorentz with some amplifications, and which attracted much attention."[1] Needless to say, Sir Edmund's slighting remark also attracted "much attention," and a good many physicists have attempted to explain why he misunderstood the situation. Of more interest is the question why, taking the paper of 1905 at face value, one has the impression that Einstein had never studied the work of his contemporaries. Given the way he was living, and his educational and professional background, it is plausible that, in fact, he had never studied most of them. Although, within a few years after his paper was published, Einstein was in correspondence with many of the leading physicists in Europe, he once remarked that he never met a real physicist until he was thirty. The only person he was able to discuss his ideas with was an engineer, Michelangelo Besso, also then an employee at the patent office, whom Einstein had known since his student days in Zurich and whom he has immortalized in the last sentence of his 1905 paper: "In conclusion I wish to say that in working at the problem here dealt with I have had the loyal assistance of my friend and colleague M. Besso, and that I am indebted to him for several valuable suggestions."

These matters can be focused sharply if one considers the physics of "time" and compares Einstein's approach to it with that of his contemporaries. The physicists of the nineteenth century inherited a vague concept of "absolute time" (and "absolute space" as well) that can be traced back to the common-sense physics of the Greeks. We need only compare Aristotle's characteriza-

---

[1] E. T. Whittaker, *A History of the Theories of Aether and Electricity* (paperback ed., New York, 1960), II, 40.

tion in *The Physics*—" 'the passage of time' is current everywhere alike and is in relation with everything"— to Newton's celebrated first *Scholium* in the *Principia*— "Absolute, true, and mathematical time, of itself, and from its own nature, flows equably without relation to anything external, and by another name is called duration. . . ." This "absolute time," of whose existence Newton is in no doubt, he compares to "relative, apparent, and common time," which "is some sensible and external (whether accurate or unequable) measure of duration by the means of motion, which is commonly used instead of true time; such as an hour, a day, a month, a year." In other words Newton attempted to distinguish between "common" time as measured by clocks and some sort of "absolute" time whose primary existence was in the consciousness of God. If pressed on this distinction, as he was in the famous correspondence between Liebniz and the Reverend Samuel Clarke, Chaplain to the Prince of Wales and a protégé of Newton, Newton retreated to a position of impregnable theological mysticism. As Clarke wrote, certainly with the approval of Newton, in his fourth reply to Liebniz, ". . . space and duration are not *hors de Dieu*, but are caused by, and are immediate and necessary consequences of His existence."[2] A modern reader, studying this correspondence, would almost certainly conclude that, from a scientific point of view, Liebniz, who took the position that absolute space and time were more or less scientifically meaningless, had gotten the better of the argument. However, for the first time in human history, a theory—Newton's—had been invented which enabled one, apparently, to calculate and predict almost everything, and it is little wonder that there is an air of impatience in Newton's attitude not unlike that of God when he replied to the not unreasonable questions of

[2] Arnold Koslow, ed., *The Changeless Order*, pp. 39, 73. This book is a useful source for historically important literature on space and time.

Job by asking, "Where wast thou when I laid the foundations of the earth?"

In the subsequent two centuries, as Newton's theory was developed and expanded, its theological substructure was forgotten and "explanation" in physics became synonymous with the reduction of physical phenomena to Newtonian mechanical models. This, as we have seen, took a dramatic turn when, following the prediction of Maxwell, Hertz demonstrated the existence of propagating electromagnetic waves in *empty space*. From the mechanical point of view, according to which every phenomenon required a mechanical model for its comprehension, this phenomenon was "incomprehensible"; hence a mechanical medium, the ether, was postulated as the seat of these vibrations. As someone remarked, " 'Ether' became the subject of the verb 'to oscillate.' " But, we have also seen, this picture came under attack when Michelson and Morley failed to detect the motion of the earth through the ether, although the theory clearly predicted that this motion should be observable under the conditions of their experiment. In order to save the ether FitzGerald and independently Lorentz were led to propose that material objects in motion contract—so that, for example, a meter stick in motion would be shortened by an amount that, in first approximation, is of the order of magnitude of the ratio of the square of the velocity of the meter stick to the speed of light. This is an incredibly small number for normal terrestrial velocities, and even for the relatively rapid motion of the earth with respect to the sun the ratio is only one part in one hundred million. Hence Lorentz and FitzGerald were not asking for much of a contraction. However, as Lorentz viewed the situation, the problem —for him, and for Poincaré—was how to *explain* this contraction in terms of some sort of model of material objects. Lorentz had indeed proposed an explanation of the contraction in terms of an electromagnetic model of matter in which material objects were taken to be

charged particles—"electrons"—situated in the ether.
These particles acted on one another with the Lorentz
electromagnetic force, and it was Lorentz's idea that
when such a system was set in motion these forces
acting in the stationary ether would be modified in such
a way as to produce the contraction.

Lorentz worked on this idea for nearly ten years and
in 1904 published a paper on "Electromagnetic Phenomena in a System Moving with Any Velocity Less Than
That of Light," which contained the most sophisticated
version of his theory. In it he made the assumption that
the electrons were charged spheres which contract into
ellipsoids when they are in motion. From this primal
contraction he sought to explain the over-all contraction
of the material objects composed of them.[3] In the
course of his argument he found it useful to introduce
a novel concept which he called "local time." As he
viewed it, the introduction of this local time was a kind
of mathematical trick for simplifying the equations—the
Maxwell equations—when doing computations for material objects in motion.[4] He did not attempt to attach

[3] An important objection to this model—made in fact by
Poincaré—was that no explanation was given of what held
the spheres together. We now know that in addition to the
electrical forces there are "nuclear forces" which are responsible for holding the atomic nucleus together. If only
electrical forces existed, the nucleus could never be stable.
[4] The transformations Lorentz introduced take the form (x,
y, and z are the position coordinates measured with respect
to three spatial axes that meet at right angles, and t is
the time):

$$x' = \frac{x - vt}{\sqrt{1 - v^2/c^2}}$$
$$y' = y$$
$$z' = z$$
$$t' = \frac{t - v/c^2\, x}{\sqrt{1 - v^2/c^2}}$$

when the transformations are to a system moving in the x
direction with a velocity v. The quantity t' is what Lorentz

any experimental meaning to the "local time." The "true" time—the only time, according to him, with physical significance—was the time as measured by an observer at rest in the ether. At first, prior to 1904, he had presented an approximate version of this transformation from the rest time to the "local time," valid for velocities that are small compared to light, but in the 1904 paper he gave an exact transformation law valid for any velocity *less* than light. These transformations play an important role in Lorentz's computation but, basically, as mathematical auxiliaries whose physical significance is obscure. It was only after Einstein's 1905 paper that the real meaning of the Lorentz transformations became clear.

The situation with Poincaré is more complicated. As one reads, for example, a lecture like "The Principles of Mathematical Physics," which Poincaré delivered in 1904 at the International Congress of Arts and Sciences in St. Louis—a year before Einstein's paper—one is continually baffled as to why Poincaré did not invent the relativity theory. In the first place, he gives a lucid statement of the "principle of relativity" itself.

called the "local time." What he showed was that the Maxwell equations have the same form in the transformed system as they do in the x, y, z, t system. This system was for Lorentz the ether absolutely at rest. He used the invariance of the Maxwell equations to simplify his calculations.

One curious feature of Lorentz's 1904 paper is that there is no explicit discussion of how his restriction to velocities less than light is to be reconciled with the Newtonian force law, F = ma, which he was using. What Lorentz *did* show was that his contracted spheres in motion increased their inertia over what it was when they were at rest, which made it more difficult to accelerate them as they went faster. As we shall see, it was Einstein's 1905 paper which cleared all of this up, since he showed that such an increase in the effective mass was a general consequence of the relativity principle and had nothing to do with any specific dynamical model of matter.

The principle of relativity [is that] according to which the laws of physical phenomena should be the same, whether for an observer fixed, or for an observer carried along in a uniform movement of translation; so that we have not and could not have any means of discerning whether or not we are carried along in such a motion.[5]

Poincaré then cites the evidence for it, in particular the work of Michelson, who "has pushed precision to its last limits."[6] However, in the very next sentence Poincaré makes clear what distinguished his thinking from that of Einstein. The principle of relativity *is to be explained*, "for which mathematicians are forced today to employ all their ingenuity."[7] This, as Poincaré viewed it, was what Lorentz was attempting to do with his dynamical explanation of the Lorentz contraction, although Poincaré indicates he felt Lorentz had used too many arbitrary hypotheses. He then discusses Lorentz's "local time" and points out that it can be understood in terms of an analysis of simultaneity much in the spirit of Einstein's. Moreover, he has a clear appreciation of the fact that since the Lorentz transformations cease to make sense when the velocities exceed that of light, a new mechanical theory—replacing Newton's—must be found in which this will be a feature. "Perhaps, likewise, we should construct a whole new mechanics of which we only succeed in catching a glimpse, where inertia increasing with the velocity, the velocity of light would become an impassable limit."[8] Two paragraphs later the lecture ends, the "new mechanics" an unrealized hope and conjecture.[9]

[5] The entire lecture is given in L. Pearce Williams, ed., *Relativity Theory: Its Origins and Impact on Modern Thought*, pp. 39–49. This passage is on p. 40.
[6] *Ibid.*, p. 41.
[7] *Ibid.*
[8] *Ibid.*, p. 48.
[9] The lasting contribution Poincaré did make to relativity theory was his discovery that the Lorentz transformations

In creating the "new mechanics" Einstein, simply speaking, managed to stand much of this on its head. In the first place, he threw out the ether altogether.

If mechanics was to be maintained as the foundation of physics, Maxwell's equations had to be interpreted mechanically [i.e., in terms of the ether]. This was zealously but fruitlessly attempted, while the equations were proving themselves fruitful in mounting degree. One got used to operating with these [electric and magnetic] fields as independent substances without finding it necessary to give one's self an account of their mechanical nature; thus mechanics as the basis of physics was being abandoned, almost unnoticeably, because its adaptability to the facts presented itself finally as hopeless.[10]

The only mention of "ether" in Einstein's 1905 paper is the famous sentence in the second paragraph: "The introduction of a 'luminiferous ether' will prove to be superfluous inasmuch as the view here to be developed will not require an 'absolutely stationary space.'" In the second place, in Einstein's formulation the principle of relativity is not something to be deduced from the theory, but is rather something that forms part of the axiomatic basis of the theory from which consequences are to be deduced. This is not to say that the principle of relativity cannot be experimentally tested. If the conclusions that are deduced from it disagree with experiment then we can conclude that the principle is wrong.

To the relativity principle Einstein adjoined a second hypothesis, namely that the speed of light in vacuum is, in his language, "a universal constant," by which he meant that it is the same no matter what speed the

form a group. This means that when two successive Lorentz transformations are combined, the resultant transformation is still a transformation of Lorentz's type. This group is now referred to as the Poincaré group.
[10] Paul Arthur Schilpp, ed., *Albert Einstein: Philosopher-Scientist*, pp. 25–26.

source of the light has with respect to the observer. (It is worth emphasizing again—since the point is often misstated—that this "constancy" of the speed of light holds only among frames of reference moving *uniformly* with respect to each other. An observer *accelerated* with respect to a light source will measure a speed of light that is different than the speed of light measured in the rest frame.) As we have already mentioned, he realized that these two hypotheses contradicted each other if Newtonian mechanics was correct. But Newtonian mechancs is built on the assumption of an absolute time. This leads to the contradiction in the following way: let us imagine a light wave propagating with the velocity of light and imagine some sort of vehicle that can propagate with, say half the velocity of light as measured with respect to the earth. Now according to Newton's laws, it follows that *viewed from the vehicle* the relative speed of light is only *half* of what it is when viewed by an observer on earth. This Newtonian "addition theorem" for velocities, one can readily show, follows from the concept of absolute time and clearly leads to a result that is in contradiction with the assumption that the velocity of light is a universal constant. In Newtonian physics by going fast enough one can always catch up with a light ray, so that the speed of light would differ for observers in different states of relative motion. To avoid this Einstein was led to challenge the epistemological basis of the assumption of an absolute time, and, as we have indicated previously, concluded that this concept was wrong. Once absolute time is abandoned there is no apparent contradiction between the two hypotheses and one is free to draw from them whatever conclusions one is led to.

One of the first consequences to emerge in Einstein's paper were the Lorentz transformation equations themselves (which he had not known of). From Einstein's point of view, the "local time" is not a mathematical trick but is inherent in the notion of time as something measured by clocks. As we have previously argued, while it is possible to synchronize clocks that are at rest with respect to each other but separated by large distances, this same procedure leads to the conclusion that moving clocks are not synchronized with respect to rest clocks, and that as seen by an observer at rest the moving clocks are "slow"—have longer periods.

The Lorentz-FitzGerald contraction also emerges from these arguments. From the point of view of Einsteinian relativity it is not needed for the explanation of the Michelson-Morley experiment, since this experiment takes place in what can be regarded, practically speaking, as a nonaccelerating frame of reference. Hence, according to the theory, no effect of the motion is to be expected, and none is found. However, the contraction would enter if we wanted to describe the experiment from the vantage point of the sun. As seen from the sun Michelson apparatus on the earth would be in uniform motion and the arm of the interferometer parallel to the direction of motion would shrink in accordance with the Lorentz-FitzGerald contraction.[11] However, in Einstein's theory, the contraction is not attributable to any particular model of the forces that hold matter together but is, rather, a feature of a careful operational definition of what is meant by "length."

The contraction of "length" is built into the procedure by which lengths are measured by rulers, just as the dilation of "time" has nothing to do with the material construction of a given clock but rather with the procedure involved in the comparison of the readings of

[11] The relativistic time dilation would also enter this description, since the moving clocks and rest clocks have different rates. One can see that without the inclusion of the time transformation the velocity of light in the two frames would differ in violation of the principle of constancy. In other words, while the Lorentz contraction alone could "explain" the Michelson experiment, it could do so only if the velocity of light was frame dependent.

clocks. To put things baldly, any model of matter consistent with Einstein's postulates *must* produce the Lorentz contraction. The Maxwell equations are invariant under the Lorentz transformations, which is the mathematically precise way of saying that they satisfy the relativity principle. Hence an electrodynamic model must have the Lorentz contraction as a feature. But so will any other model that is relativistically invariant. The Lorentz contraction is difficult to check by direct experiment, but indirect checks abound. In particular, since velocity is the ratio of distance to time and since both distance and time are affected by the Lorentz transformation, one would expect that velocity would also be affected. This is the case, and Newton's addition theorem for velocities is modified in such a way that any velocity added to the velocity of light is still the velocity of light, which is another way of saying that a material object cannot overtake a light ray.[12]

[12] In Newtonian mechanics the relative speed w of the material object is given by:

$$w = c - v$$

while in relativity theory it is given by:

$$w = \frac{c - v}{1 - v/c} = c$$

which means that as seen from the moving vehicle the velocity of light is always c.

Einstein's original discussion of the Lorentz contraction involved an idealized ruler whose actual physical dimension was assumed to be negligible—an arbitrarily thin ruler. In 1959 a physicist named J. Terrell wrote a paper on "The Invisibility of Lorentz Contraction" (*Physical Review*, 116 [1959], 1041), in which he raised the question of what would one actually *see* if, say, a cube were moved toward an observer at a speed close to that of light. By an intricate argument involving relativistic optics he reached the conclusion that the cube would appear not *flattened* but *rotated*, if, say, it was viewed from the side. This does not contradict the Lorentz contraction, which actually enters Terrell's argument, but is a consequence of the fact that light emitted simultaneously from various parts of an object of finite size takes unequal amounts of time to reach an observer,

In Einstein's paper these results are produced by simple, general arguments, as opposed to long calculations of the type that Einstein's contemporaries were accustomed to. This is what made it so difficult for most of them to understand, at first, what he was doing. Not only had he been led to new results, but also he had introduced a new way of thinking about physical problems.

---

since each light beam travels a different distance. When these different light beams are integrated to give an image, the relativistic cube appears rotated and cubical.

$E = mc^2$

viii

One of the most significant aspects of Einstein's
1905 paper is the natural unification it gives to
the concepts of electricity and magnetism. This
unification is contained in the Maxwell equations, but the relativity theory gives a novel way
of looking at it. This is most clearly illustrated
if we consider a single electron. If we are at
rest with respect to this electron, it produces
a pure electric force. We can test this by taking
another electron initially at rest with respect to
the first and observing that these two particles
repel each other—like charges always repel, according to Coulomb's law, named after the
French physicist who discovered it in 1785. (The
law closely resembles Newton's law of gravitation in that the force depends on the product
of the electric charges divided by the square of
the distance between them, whereas Newton's

law says that the gravitational force, which is always attractive, depends on the product of the *masses* divided by the square of the distance between *them*.) However, once the electron is set into motion the electric force is modified, and in addition the electron generates a magnetic force. This was known before the relativity theory, but in his paper Einstein demonstrates how these two situations are simply related to each other by a Lorentz transformation. In other words electricity and magnetism are essentially the *same phenomenon*—or are different aspects of the same phenomenon—and which aspect is emphasized will depend on the velocity of the observer with respect to the electron. This had been partially appreciated in various ways, based on the Maxwell equations and special models for electrons, prior to Einstein, but it was only with Einstein's theory that the connection between electricity and magnetism was fully understood.

In fact, Einstein realized something else which turns out to be of great importance in testing the theory. Put simply, a moving electron, or any massive object, becomes more massive when it is in motion with respect to an observer than when it is at rest with respect to the same observer. In particular, as the speed of the object approaches that of light its mass becomes *infinite*! No doubt this seems very strange, although after what has gone before the reader may be prepared to believe anything. The common-sense idea of mass is that it represents the quantity of matter in an object. How can the quantity of matter change simply by putting the object in motion? But the "quantity of matter" is not exactly what the physicist means by "mass." To see what the physicist means imagine that we have two objects—a billiard ball and a bowling ball, for example—and subject them to identical forces. (We assume that we understand enough of the nature of the force in question so that we can be sure that the two forces really are identical.) According to Newton's law each object will begin to accelerate. But these accelerations will be different. The billiard ball is easier to accelerate than the bowling ball. This is because the bowling ball is more "massive." Precisely speaking, we define the "inertial mass"—later we shall be led to discuss another concept of mass, the "gravitational mass," which measures the response of objects specifically to the force of gravity— as the ratio of the force acting on an object to the acceleration it produces. This states in a precise way the idea that a bowling ball is more difficult to accelerate than a billiard ball because it is more massive. Hence when a physicist says that the inertial mass of an object increases with its velocity, what he means is that the same force applied to it will have less and less success in accelerating it as the object moves faster and faster, and as the object approaches the speed of light it will have no success at all. Before Einstein various physicists had conjectured that something like this effect existed for electrons, on the basis of specific calculations with the Maxwell equations and various models of the electron. In fact in 1901 the German physicist Walter Kaufmann began a long series of measurements to determine if electrons showed an increase in inertial mass as they are accelerated and, if so, by how much. He used electrons emitted in the radioactive decay of radium. (Radioactive decay of uranium into charged particles had been discovered by the French physicist A. H. Becquerel in 1896, and for a time these particles were called "Becquerel rays" until they were later identified as electrons—the same charged objects that were known to boil off heated metal plates.) His results agreed with some theories and disagreed with others. In particular, they disagreed with the predictions of Lorentz's paper of 1904 in which the electrons were taken as tiny charged spheres that contracted into ellipsoids when in motion. More significantly, they disagreed with the results of Einstein's 1905 paper in which he had obtained Lorentz's answer as a straightforward prediction of the relativity theory and independent of *any* special model of the electron. As Gerald Holton points out, the first reference to Einstein's 1905 paper in the physics literature is to be found in a paper by Kaufmann published in the *Annalen der Physik* in 1906, which begins with the summary statement, "I anticipate right here the general result of the measurements to be described in the following: *the measurement results are not compatible with the Lorentz-Einsteinian fundamental assumptions*" [Italics in the original.]

Holton goes on to add:

> Einstein could not have known that Kaufmann's equipment was inadequate [i.e., that his experiment was wrong]. Indeed, it took ten years for this to be fully realized. . . . So in his discussion of 1907 [a summary article on relativity for the *Jahrbuch der Radioactivitat und Elektronik*] Einstein had to acknowledge that there seemed to be small but significant difference between Kaufmann's results and Einstein's predictions. He agreed that Kaufmann's calculations seemed to be free of error but "whether there is an unsuspected systematic error or whether the foundations of relativity theory do not correspond with the facts one will be able to decide with certainty only if a great variety of observational material is at hand."

Now—and this is the significant part, as it revealed Einstein's extraordinary scientific attitude—Holton notes:

> Despite this prophetic remark, Einstein does not rest his case on it. On the contrary, he has a very different, and what for his time and situation must have been a very daring point to make: He acknowledges that the theories of electron motion given earlier by Abraham and by Bucherer do give predictions considerably closer to the experimental results of Kaufmann. But Einstein refuses to let the "facts" decide the matter: "In my opinion both theories have a rather

small probability because their fundamental assumptions concerning the mass of moving electrons are not explainable in terms of theoretical systems which embrace a greater complex of phenomena." . . .

This is the characteristic position—the crucial difference between Einstein and those who make the correspondence with experimental fact the chief deciding factor for or against a theory: Even though the "experimental facts" at that time very clearly seemed to favor the theory of his opponents rather than his own, he finds the *ad hoc* character of their theories more significant and objectionable than an apparent disagreement between his theory and their "facts."[1]

Such an attitude—the disregard of some apparent experimental facts because they do not fit into a wide-ranging and harmonious theoretical pattern—is either an attribute of sheer genius or pure folly, depending on who holds it and for how long. The young Einstein seemed to have an intuition for scientific truth so powerful that he seemed incapable of being wrong. When Einstein turned the full force of this same intuition, in middle age, against the quantum theory, many physicists would argue that it did, perhaps, border on folly.

The reader has by now perhaps noticed that there has been no mention of the most celebrated result of Einstein's relativity theory, the formula which has, alas, become almost synonomous with his name: $E = mc^2$. The reason is that this formula, with its correct interpretation, is not in his 1905 paper, although it is there implicitly. The formula with its correct interpretation emerged in a remarkable three-page paper that Einstein published, also in 1905 and also in the *Annalen*, entitled "Ist die Trägheit eines Körpers von seinem Energienhalt abhängig?" ("Does the Inertia of a Body Depend upon Its Energy Content?")

[1] Gerald Holton, "Mach, Einstein and the Search for Reality," pp. 651–52.

This paper is a perfect model for what the deductive process in physics is at its best. Like the rest of Einstein's early papers it is almost completely non-mathematical, but rests on inventing a "thought experiment" which when carefully analyzed yields the result. In this case, Einstein imagined an atom, or some other particle, that decayed radioactively by emitting light radiation—gamma rays.[2] We now know many examples of such decays, but in 1905 the study of radioactivity was still in its infancy and decays such as Einstein imagined had not yet been studied in full detail. By applying the principle that energy and momentum had to be conserved in the decay and by making an ingenious use of the Lorentz transformation, he was able to argue that the atom that was left after the decay had to be less massive than the original atom. Furthermore, the amount of mass that was lost was just equal to the total energy, E, carried away by the radiation, divided by the square of the velocity of light; or, in formulas, if m is the mass loss, then $m = E/c^2$. In his words, *"If a body gives off the energy* E *in the form of radiation its mass diminishes by* E/c^2.*"*[3] (The italics are in the original, where the symbol "L" is used instead of "E" for "energy.") This is all there is to it, except for a penultimate sentence in his note which must rank as one of the most extraordinary understatements ever published in a physics paper. "It is not impossible that with bodies whose energy-content is variable to a high degree (e.g. with radium salts) the theory may be successfully put to the test."

Fundamentally, the Einstein equation revealed a new and until then unsuspected source of energy. The mere

[2] Symbolically the decay can be represented as:

$$A \rightarrow B + \gamma + \gamma$$

where A is the initial decaying particle, B is the final particle, and γ stands for a quantum of light.

[3] *The Principle of Relativity*, p. 71.

fact that a material object has mass endows it with an energy, mc², which is very substantial because the velocity of light is so large. In general, this energy is unavailable in a practical sense, which is why it had not been discovered prior to Einstein, since it cannot be readily converted into the forms of energy—light, heat and so on—that we *can* make use of. When a radio-active particle decays, say, into light radiation, this conversion is made naturally—some mass energy becomes converted into radiation energy. This was what Einstein had in mind in writing the sentence quoted above. We now know that there are extreme examples involving some of the more recently discovered "elementary particles" where this conversion of mass into radiant energy is complete.[4] The original particle disintegrates spontaneously and all its mass-energy becomes converted into the energy of radiation. In these cases one finds a certain number of gamma rays—very energetic light quanta—whose total energy adds up to the mass-energy of the particle that has disintegrated. Another interesting case is one in which two massive particles come together and annihilate each other completely, yielding light quanta. The first such example to be studied was the mutual annihilation of the electron and its "anti-particle"—the positron. The positron was discovered in cosmic rays in 1932 by the American physicist C. D. Anderson (who was awarded the Nobel Prize in 1936). Its existence had been predicted by the British theoretical physicist P. A. M. Dirac in 1930, who made the first honest marriage between the relativity theory and the quantum theory and discovered that one of the inevitable offspring of this union were anti-particles—objects that have the same mass as the particle, but, if electrically charged, have the opposite

---

[4] An example is the decay of the so-called neutral π meson —the $\pi°$. The $\pi°$ has a rest energy of approximately 140 million electron volts. In the decay $\pi° \rightarrow \gamma + \gamma$ each light quantum takes away half of the $\pi°$'s rest energy.

charge. All particles have anti-particles, but the electron-positron pair was the first such pair to be discovered. When an electron and positron meet each other, at rest, they can annihilate into two light quanta, each of which carries away an energy equal to mc^2 where m, in this case, is the mass of the electron (or positron).

Another very important application of Einstein's result is in "nuclear fusion," which may one day become our most important source of cheap, pollution-free power. To begin with the simplest example, a neutron and a proton can join to form a deuteron—the nucleus of "heavy hydrogen"—losing a certain amount of energy, which is carried off by a light quantum. This energy loss means, according to Einstein's principle, that the deuteron must have less mass than the sum of its parts—the neutron and proton masses. This is, indeed, the case, and the mass difference is called the "binding energy" of the deuteron, since to break the deuteron up again we must supply at least this energy. (The deuteron can be broken up into a neutron and proton by applying light quanta of suitable energy, a process called "photodistintegration.") One can imagine constructing all the stable nuclei by fusing together neutrons and protons, and it is clear, from what has been said, that all these nuclei will have masses that are less than the sum of the masses of the constituent parts, i.e., less than the sum of the neutron and proton masses. There was no mechanism known to classical physics that could account for such mass losses. Einstein's formula gives a simple and natural explanation, since in the formation of such nuclei by fusion some energy—or mass—is given up and hence the resulting nucleus is *less* massive than its constituent parts.

An important problem that was resolved by nuclear fusion and the energy generated therefrom was how the stars, including the sun, continued to "burn"—i.e., emit such large quantities of radiant energy—for so long and

with such intensity. The first and most naïve explanation
offered was that the sun is a lump of coal set on fire.
This happy idea is instantly refuted if one computes
how long it would take such a lump to burn itself out
at the observed rate at which the sun produces energy.
The answer is about 1500 years, whereas the age of the
solar system is several billion years. The next most
sophisticated idea was put forth in 1854 by the German
physicist Hermann Helmholtz. Helmholtz made use of
the notion, which is still accepted, that the sun was
formed out of a gaseous cloud when the gas particles,
for reasons that are still being debated by astronomers,
began to cluster in lumps. These primal clouds pre-
sumably consist mainly of hydrogen and helium, with a
small admixture of heavy elements such as iron—the
so-called "cosmic dust." Once this process starts, the
mutual gravitational attraction among the particles will
keep it going and the excess energy that is available
from gravity as the star collapses can be converted into
radiation. However, it was soon shown that if this were
the *only* process involved the sun could not be older
than some 20 million years. After the discovery of radioactivity the idea was advanced that perhaps the sun's
energy is due to radioactive decay. If the sun were made
of pure uranium it would give off energy of something
like the observed rate over several billions of years. But
the sun is *not* made of uranium but, rather, of light
elements like hydrogen and helium. This left things
pretty much at an impasse until the 1920s, when the
late George Gamow, using ideas of the then newly
discovered quantum mechanics, suggested that the
fusion process could actually take place at the temperatures in the interior of stars. (It had been clear to and
clearly stated by the British astronomer Sir Arthur
Eddington for several years prior that Einstein's formula must be the key to the situation, but the actual
mechanism of fusion required quantum mechanics for

its quantitative understanding.) In 1939 H. A. Bethe and C. F. von Weizsäcker independently gave the detailed nuclear chemistry—the specific reactions—that take place in this nuclear cooking and hence, pretty much, completed the explanation of the generation of solar energy. (There is in this picture no problem with the long lifetime of the sun, since so little mass loss produces so much energy.) Thus Einstein's three-page paper of 1905 has been among the most fruitful in modern physics.

All these experimental confirmations came much later, and the initial reaction to Einstein's 1905 papers on relativity was, in the main, indifferent or negative. Infeld writes:

What was the impact of these new ideas? At first there was hardly any. Nowadays important results are perhaps more quickly recognized and a revolutionary new paper often produces a flood of other contributions, written by men who build up the ideas in greater detail and who develop them mathematically. But such a flood of papers did not occur immediately after the appearance of Einstein's articles. It began some four years later—a long time interval for scientific recognition. Yet I know that there were physicists who read Einstein's paper very carefully in the interim and who saw in it the birth of a new science. My friend Professor Loria told me how his teacher, Professor Witkowski in Craeow (and a very great teacher he was!), read Einstein's paper and exclaimed to Loria, "A new Copernicus has been born! Read Einstein's paper." Later, when Professor Loria met Professor Max Born at a physics meeting, he told him about Einstein and asked Born if he had read the paper. It turned out that neither Born nor anyone else there had heard about Einstein. They went to the library, took from the bookshelves the seventeenth volume of *Annalen der Physik* and started to read Einstein's article. Immediately Max Born recognized its greatness and also the necessity for formal generalizations. Later, Born's own work on relativity theory became one of the most important early contributions to this field of science.[5]

Of the men we have discussed, Poincaré began to have doubts about the relativity principle itself after the early experimental results of Kaufmann on the mass of the electron, and in 1906 he wrote that these "experiments have given ground to the Abraham theory [a model of the electron which differed from that of Lorentz]. The principle of relativity may well not have the rigorous value which has been attributed to it."[6] Poincaré died in 1912 before the matter was completely settled. From 1905 until his death, Poincaré wrote and lectured frequently about the relativity principle. But in none of these lectures did he ever mention Einstein's contribution, although he described Einstein's work in other branches of physics.[7] However, in 1911, the year before he died, Poincaré wrote a letter of recommendation for Einstein, who was then being considered for a job at the Polytechnic in Zurich, in which he said:

Einstein is one of the most original minds that I have known; despite his youth he has already achieved a very honorable rank among the foremost scholars of our time. What we can, above all, admire in him is the facility with which he adapts himself to new concepts and draws all the consequences from them. He does not remain attached to classical princi-

[5] Leopold Infeld, *Albert Einstein, His Work and Influence on Our World*, p. 44.
[6] Quoted in Stanley Goldberg, "Henri Poincaré and Einstein's Theory of Relativity," p. 938.
[7] Poincaré left no private papers, and so his reasons for publicly ignoring Einstein's contributions to the relativity theory are not understood. See Stanley Goldberg, "In Defense of Ether," in Russell McCormach, ed., *Historical Studies in the Physical Sciences 1970*, pp. 88–125, for a discussion of this and an analysis of the reactions of European physicists to the relativity theory.

ples and in the presence of a physics problem is prompt to realize all of its possibilities. This translates itself immediately in his mind by the prediction of new phenomena, which can be verified by experiments. I do not mean that all of his predictions will be confirmed when they are eventually tested. Since he searches in all possible directions, one should, on the contrary, expect that most of the paths that he follows will lead to an impasse; but one may also hope that one of the directions that he has pointed out will be the true one; and that is enough. This is the way that one ought to proceed. The role of mathematical physics is to ask questions; it is only experience that can answer them. The future will show, more and more, the worth of Einstein, and the university which is able to capture this young master is certain of gaining much honor from this operation.[8]

Lorentz's acceptance of the relativity theory increased progressively and, in 1915, he added a footnote to his classic text *The Theory of Electrons*:

If I had to write the last chapter now, I would certainly have given a more prominent place to Einstein's theory of relativity by which the theory of electromagnetic phenomena in moving systems gains a simplicity that I had not been able to attain. The chief cause of my failure was my clinging to the idea that the variable t [the time as measured in the "ether" frame of reference] can only be considered as the true time and that my local time t′ must be regarded as no more than an auxiliary mathematical quantity. In Einstein's theory, on the contrary, t′ plays the same part as t; if we want to describe phenomena in terms of x′, y′, z′, t′ we must work with these variables exactly as we could do with x, y, x, t.[9]

[8] A copy of the original French text was kindly supplied by Helen Dukas. It was first published in Carl Seelig's *Albert Einstein* (Zurich: Europa Verlag, 1960), pp. 228–29.
[9] H. A. Lorentz, *The Theory of Electrons*, p. 321.

When Lorentz wrote this footnote he was sixty-two. He had come to the small university town of Leyden in Holland as a very young man and he remained there until his death in 1928. Perhaps it was a question of age. Both Lorentz and Poincaré were in their late forties when they were responding to the crisis that they felt had been produced in physics by the Michelson-Morley experiment. In a sense they knew too much and had too much of a vested interest in classical physics to throw it away. Einstein was only twenty-six when he published his paper, and indeed, not only his but all the really great discoveries in theoretical physics—with a few exceptions that stand out because of their oddity —have been made by men under thirty. And in many cases the older generation did not, or could not, react with the character and flexibility of Lorentz. While Lorentz continued to refer to the ether, even to the end of his life, as a "conception" that "has certain advantages,"[10] it was he who sent Einstein the first telegram in September 1919 announcing that the British solar eclipse expeditions had confirmed the predictions of Einstein's general theory of relativity. The two men, despite the vast difference in their ages, backgrounds, and temperaments, developed a friendship and deep mutual admiration which lasted to the end of Lorentz's life and which Einstein cherished until the end of his own life. Albert Michelson, on the other hand, for whom Einstein had an enormous respect as an experimenter, disliked and disbelieved the theory until the end. When

[10] Something of the ambiguity in Lorentz's mind can be seen in the full quotation: "As to the ether . . . though the conception of it has certain advantages, it must be admitted that if Einstein had maintained it he certainly would not have given us his theory, and so we are grateful to him for not having gone along old-fashioned roads." H. A. Lorentz, *Problems of Modern Physics: A Course of Lectures Delivered in the California Institute of Technology* [1922], H. Bateman, ed. (Boston 1927), pp. 220–21.

they met for the first and only time—in 1931, when Michelson was seventy-nine—Michelson remarked to Einstein that he was even a little sorry that his experiments had been responsible for giving birth to such a "monster."[11]

[11] See Gerald Holton, "Einstein and the 'Crucial' Experiment," p. 973.

Four-Dimensional Space-Time

iX

The steps that led Einstein from Berne to Berlin
began when Swiss physicists came to realize his
talents were being wasted in a patent office.
"The researches whose results Einstein published
at Bern in 1905 were so unusual that to the
physicists of the Swiss universities they seemed
incompatible with the assigned work of a minor
official of the patent office," Philipp Frank has
written, and "attempts were soon made to bring
Einstein to teach at the University of Zurich."[1]

Here Einstein and Professor Alfred Kleiner,
the leading physicist of Zurich and Einstein's
principal advocate there, ran into a snag, since
according to the rules no one could become an
appointed professor before having served for some
time as a *Privatdozent*. This was a strange academic position whose holder had no obligations
and the university had no obligations to pay him

[1] Philipp Frank, *Einstein: His Life and Times*, p. 74.

either. The *Privatdozent* could give what lectures he felt like, for which the students paid a small fee. It was almost impossible to make a living at this, and so Einstein also retained his job at the patent office. But despite everything Einstein finally was called to Zurich in 1909 as a professor "extraordinary," which, despite the grandiose title, was an undistinguished academic job. In fact his salary remained what it had been at the patent office, while he had to give up the relatively inexpensive life he was leading in Bern, since a university professor had a certain unavoidable minimum of social responsibility.

Einstein always retained a great fondness for Zurich, but in the fall of 1910 there was a vacancy in the chair of theoretical physics at the German University in Prague, which meant a promotion. Until 1888 there had been one university in Prague, the oldest in central Europe, with professors lecturing in both German and Czech. However, after incessant political quarreling between Germans and Czechs, the Austrian government decided to split the university in two, by language. This led to even more hostility, since the German-speaking group looked down on the Czechs as "racially inferior." It was a requirement, imposed by the Emperor Franz Josef, that only members of a recognized church should teach in the state universities, so when he came to Prague in 1911 Einstein, who had not practiced any formal religion since the age of twelve, declared that he belonged to the "Mosaic" creed, which was the official designation for Judaism in Austria at that time. Einstein had a fair amount of contact with the Jewish literary circle in Prague, which included Kafka, Hugo Bergmann, and Max Brod, who, four years later in a novel, *The Redemption of Tycho Brahe*, portrayed the character of Johann Kepler, Tycho's assistant and scientific antagonist, in a way that was clearly based on the character of the young Einstein. In 1912 Einstein left Prague and returned to the Polytechnic in Zurich as a professor. Throughout this period he was working intensively in various branches of physics. Here we wish to focus on the development of relativity theory. Later we shall return to quantum physics, to which Einstein was simultaneously making crucial and fundamental contributions.

Curiously, the next significant development in relativity came not from Einstein, but from his old teacher in the Polytechnic in Zurich, Hermann Minkowski, a Russian-German mathematician who was born in 1864 and died in 1909 two years after he did his fundamental work. Minkowski, who had a generally unfavorable recollection of Einstein from his classes, had moved to the University of Göttingen, which was at that time and for many years after the mathematics capital of the world. It is Minkowski's formulation of the special theory of relativity that, in the main, we now teach, because of its formal simplicity and mathematical elegance. At this period Einstein was still allergic to pure mathematics, and for several years he was not particularly enthusiastic about Minkowski's "four-dimensional world view." It was only when he found the final formulation of his theory of gravitation, a sweeping generalization of Minkowski's work, that he fully appreciated its formal power. A bit of Einstein's earlier attitude remains in the introductory sentence of the chapter entitled "Minkowski's Four-Dimensional Space" in the beautiful popular book, *Relativity*, which he wrote in 1916. "The non-mathematician," he begins, "is seized by a mysterious shuddering when he hears of 'four-dimensional' things, by a feeling not unlike that awakened by thoughts of the occult." But then he adds, "And yet there is no more commonplace statement than that the world in which we live is a four-dimensional space-time continuum."[2] Whether or not this is a "commonplace statement" depends, to some extent, on the circles one moves in. What Einstein no doubt means is that we unconsciously adopt a four-dimensional de-

[2] *Relativity*, p. 55.

scription of events without being aware, unless we happen to be mathematicians, that this is what we are doing. When we agree to meet someone at a given location at a given time we are making a four-dimensional statement since the location, can, precisely speaking, be specified by three spatial coordinates, x, y, z, plus the time, t, the fourth coordinate or "dimension." Hence the over-all meeting place in "space-time" is specified by what mathematicians call a four-dimensional vector. None of this has anything to do with the relativity theory, and Newtonian mechanics can perfectly well be formulated in four dimensions, except that it is not interesting to do so. The reason is that in Newtonian physics time is "absolute" and its measure does not depend on which uniformly moving frame of reference—which "inertial frame"—we use in our description. Hence, the transformation equations that take one from one inertial frame to another in Newtonian physics are only "interesting" for the three spatial dimensions. The fourth equation simply says that "time" in one frame is identical to "time" as it is measured with respect to any other inertial frame. In the Lorentz transformations of the special theory of relativity *both* space and time transform, so in this sense they are on the same footing and it is natural to treat them conjointly in four dimensions. This is, in fact, what Einstein did implicitly in his 1905 paper. However, Minkowski showed how one could "visualize" these transformations in a precise geometrical way.

As everyone knows, the celebrated Pythagorean theorem of Euclidean geometry states that if we are given a right triangle—a triangle with a 90° angle in it—the sum of the squares of the lengths of the two sides equals the square of the hypotenuse. We can make use of this theorem in the following way. First we set up a pair of "axes"—two lines that meet at right angles at a point that we call the "origin." Now if we draw *any* line segment that passes through the origin we can find its

length by the following device: We can drop perpendicular projections of our line on the two axes—call them
projections x and y. Our line segment is now the hypotenuse of a right triangle whose sides are x and y. Hence
the sum of these squares, $x^2 + y^2$, will give by the
Pythagorean theorem the length of our line segment.
Such a line segment is called by mathematicians a "vector," and x and y are called the "components" of the
vector with respect to the axes we have chosen.

There is clearly something arbitrary about this procedure—namely, our choice of axes. If we rotate the two
axes rigidly through some angle, leaving the vector
alone, we will obtain a new set of axes and a new set of
"components" x' and y'. However, no such rotation can
ever change the length of the vector. Hence the sum of
the squares of x' and y' are equal to the sum of the
squares of x and y. This sum is "invariant" against rigid
rotations of the axes. We can easily generalize this procedure to three dimensions by introducing a third axis
at right angles to the other two. Hence the three-
dimensional vector will be characterized by the three
components x, y, and z, and the length of this vector is
just the sum of the squares of the three components.
This length is also invariant against rotations in three
dimensions.

Now, nothing stops us, at least in our imagination,
from adding a fourth axis at right angles to the other
three, and defining a four-dimensional vector. We call
the "Euclidean length" of this vector the sum of the
squares of the four components, and this will also be
invariant against rigid rotations in four dimensions.
(This may be hard to visualize, but that is beside the
point.) As we have seen, an "event" in space-time can
be described by the four components—x, y, z, and t.
Since x, y, and z have dimensions of "length" and t has
dimensions of "time"—"meters" in one case and "seconds" in the other—we unify the description by introducing ct, instead of t, where c is the velocity of light in

vacuum. Now all four coordinates x, y, z, and ct have the same physical dimensions—meters—since the product of a velocity and a time has dimensions of length. At first sight, we might be tempted to call the length of an "event" the sum of the squares of these four components. This Euclidean length would be invariant against rigid rotations of the axes. But relativity theory teaches us that the interesting transformations are not ordinary rotations in four dimensions but are, rather, the Lorentz transformations. Now it turns out (and this was implied by Einstein in a footnote in his 1905 paper but not emphasized) that the Lorentz transformations leave invariant the quantity $s^2 = x^2 + y^2 + z^2 - c^2 t^2$. The reader will notice the fact that the space and time components have opposite signs. Hence this "length" is not necessarily positive. It can be positive if the square of the time component is less than the sum of the squares of the space components. It can be negative if the opposite is true, or it can be zero. This last, Einstein pointed out, is just the equation of motion of a light ray in vacuum. The fact that this equation has the same form in every frame of reference connected by Lorentz transformations is only an aspect of the relativity principle, which is what Einstein pointed out in his footnote. However, Minkowski went one step further in clarifying the geometry of the relativity theory. He argued that the Lorentz transformations act like special rotations in the four-dimensional space. These rotations are through what the mathematicians call "imaginary" angles, which means that unlike the ordinary angles which run from zero to 360 degrees, these angles contain the square root of minus one.[3] In other words,

---

[3] This is one of many cases encountered in the verbal description of mathematical ideas in which the verbiage is more complicated than the idea being expressed. Minkowski observed that if one writes

$$x_4 = i\,ct$$

where $i = \sqrt{-1}$ so that $i^2 = -1$, then the Lorentz trans-one has what is known as a "pseudo-Euclidean" geometry in Minkowski's four-dimensional space: "pseudo"-Euclidean because the "length" is not always positive. Minkowski introduced a rather nice name—"world line" —for the trajectories of particles in four dimensions: every point on a world line is labeled by three space and one time coordinates, so that one can follow the world history of a particle in space and time if one knows its world line. The great virtue of Minkowski's approach is that, with a little practice, it becomes easy to visualize the results of the Lorentz transformations and to tell very quickly how quantities like electric and magnetic fields will transform when one goes from one inertial frame to another. There is really no new physics in it, but the formalism is much more elegant and compact than the one in Einstein's original paper.

In 1908 Minkowski gave what must have been a rather ebullient lecture at the eightieth Assembly of German Natural Scientists and Physicians at Cologne, a lecture he called "Space and Time." One can almost say that it was this lecture that put relativity theory on the map. He begins with a statement that has been widely quoted, and misunderstood, ever since: "The views of space and time which I wish to lay before you have sprung from the soil of experimental physics and therein lies their strength. They are radical. Henceforth space by itself, and time by itself, are doomed to fade away into mere shadows, and only a kind of union

---

formation can be written as a "rotation" through an angle ψ as follows:

$$x' = x \cos \psi - x_4 \sin \psi$$
$$x'_4 = x \sin \psi + x_4 \cos \psi$$

where

$$\frac{v}{c} = i \tan \psi.$$

This has the same form as the equation of an ordinary rotation except that here ψ is an "imaginary" angle.

of the two will preserve an independent reality."[4] This has been interpreted to mean, by some science-fiction writers, poets, novelists, and philosophers, who seem to have stopped reading the lecture at this point, that somehow because of the four-dimensional aspect of relativity one can move oneself back and forth in time into the future and past and God knows what. Unfortunately, nothing like this is true. We are each attached to our own proper Lorentz frames, and for us, so far as the theory of relativity is concerned, the future remains the future and the past the past. A somewhat more subtle question arises if one asks whether causal sequences will appear the same to all observers connected by Lorentz transformations; that is, if in my Lorentz frame event B follows event A, will the observers in all other Lorentz frames see B follow A even though space and time get transformed by a Lorentz transformation connecting the various frames? The answer is yes, and one can prove from the Lorentz transformations that if, in one frame, two events at a point occur in a given time order —the second, say, later than the first—then they will appear in the same time order to all observers connected by Lorentz transformations. This is also true for events at different space points, provided that these points are such that a light signal can be exchanged between them. This question has recently come up in connection with a novel speculation originated independently by the physicists George Sudarshan and Gerald Feinberg. They have looked into the theoretical possibility that there might be particles that *always* travel faster than light. We have seen that it is impossible to accelerate an ordinary particle to a speed greater than light, but the question raised here is whether there might not be a new class of particles— which Feinberg has christened "tachyons"—that can

[4] The full lecture is reprinted in *The Principle of Relativity*. This passage is found on page 75.

never be decelerated to speeds less than light. If so, could one use these hypothetical faster-than-light objects to violate our ideas about causal sequence? It appears that this can be avoided, but it is not clear if the theory can be made fully consistent. It has been suggested how one might look for the tachyons in experiments but, so far, the tachyons have not shown up. One is inevitably reminded of the famous limerick—

> There was a young lady named Bright
> Whose speed was far faster than light;
> She set out one day,
> In a relative way,
> And returned home the previous night.

What Einstein would have thought of the tachyons we will unfortunately never know, but his initial reaction to Minkowski's papers was not enthusiastic, since he felt that what appeared to him to be an unnecessarily elaborate mathematical formalism hid the physics. When the distinguished German physicist Max von Laue, who won the Nobel Prize in physics in 1914, published the first detailed mathematical textbook on the special theory of relativity in 1911, Einstein commented jokingly, "I myself can hardly understand Laue's book." Indeed, David Hilbert, a professor at Göttingen who was probably the greatest mathematician of that era, once remarked, "Every boy in the streets of our mathematical Göttingen understands more about four-dimensional geometry than Einstein. Yet, despite that, Einstein did the work and not the mathematicians." Hilbert asked a group of mathematicians, "Do you know why Einstein said the most original and profound things about space and time that have been said in our generation? Because he had learned nothing about all the philosophy and mathematics of time and space."[5]

[5] See Frank, *op. cit.*, p. 206.

Mach's Principles

The last paper that Einstein wrote on relativity that is mathematically "simple"—*On the Influence of Gravitation on the Propagation of Light*— written in 1911, is a mid-station between the papers of 1905 on the special theory and his final formulation of the general theory of relativity: his masterpiece of 1916.[1] No one would argue that "every boy in the street," even in Göttingen, can readily follow the "four-dimensional geometry" of Einstein's 1916 paper. An average advanced student in physics reading it now is likely to have more background in Riemannian geometry, tensor calculus, Christoffel symbols, and all the rest of the mathematical bestiary that Einstein was forced to borrow or invent to deal with the problem of gravitation than had most of the physicists of that period. Even so, it is still a serious and difficult paper to

[1] A translation of the 1916 paper, "The Foundations of the General Relativity Theory," can be found in *The Principle of Relativity*, pp. 111–64.

read in detail. For many of Einstein's contemporaries it was essentially impossible. On one occasion, Sir Arthur Eddington—who was closely associated with both the theoretical and experimental development of the theory and one of the first to appreciate it—was asked if it was true that only three people in the world understood the general theory of relativity. Eddington is said to have answered, "Who is the third?"[2]

Gravitation is the weakest force (under normal conditions) that we know in nature. Suppose one compares the electric force that repels, say, one positively charged proton from another, with the force of gravity that tends to attract them together. (In making this comparison we are assuming Coulomb's law for the electric force and Newton's law of gravitation for the gravity. Neither law is, according to modern ideas, strictly correct, but they are both accurate enough for purposes of this argument.) It turns out that the electric force is stronger than the gravitational by a factor of about 10^{36}. (In the same notation a million is represented by 10^6.) In other words, in considering the physics of protons, under normal circumstances (some newly discovered astrophysical phenomena such as pulsars are an example of "nonnormal circumstances") we can, to an extremely high degree of accuracy, leave gravitation out altogether. One may then well ask why the motions of the planets are governed by gravitation and not by electricity. The straightforward answer is that planets are, essentially, electrically "neutral." This is not quite so tautological as it seems, and it points up a profound differ-

[2] Eugene Guth, "Contribution to the History of Einstein's Geometry as a Branch of Physics," *Proceedings of the Relativity Conference in the Midwest 1969* (New York, 1970), p. 185. Guth gives an interesting account of the history of general relativity. Professor S. Chandrasekhar, who knew Eddington well, has informed me that he heard this story from Eddington himself.

ence between electricity and gravity. As we have mentioned before, electric charges can be positive, negative, or neutral (as much negative as positive). It is a matter of historical convention which charge one gives the name "positive" or "negative." The important thing is that it is an experimental fact that the electron and proton have equal and opposite charges—by convention the charge of the electron is taken "negative"; and that the neutron—the third component of normal matter—has *no* net charge. This means that a normal atom, with all its electrons, is electrically neutral; hence planets, as a whole, are very nearly electrically neutral. (The two protons, in the example above, are both positively charged, which is why they repel.) And electrically neutral objects have, to first approximation, no electrical effect on each other. Hence in the theory of planetary motion electricity plays no role, although it is a much stronger force than gravity when charged particles are involved. However, what determines the strength of the gravitational force is the "mass" of the particle. Indeed, according to Newton's law of universal gravitation, any particle, whatever its electrical charge, is attracted to any other particle in the universe with a strength—or "coupling constant," to employ the modern term—that is proportional to the product of the two masses. The mass of an individual proton is infinitesimal—about one 10^{24}th of a gram (the neutron has almost the same mass)—but there are a huge number of them in, say, the sun. (A more exact figure is about 10^{57} protons.) All these tiny masses add up to produce the net gravitational attraction that the sun exercises on the planets.

It is important to notice that the term "mass" has been used in our discussion in two senses that are not manifestly the same. On one hand, we have called the "inertial mass" that property of an object which measures its response to a given force according to Newton's law $F = ma$: force equals the product of the "inertial

mass" times the acceleration produced. This mass can
be measured in experiments that have nothing to do
with gravity. For example, the inertial mass of a proton
can be measured by its response to an electric force. On
the other hand, there is the "gravitational mass." This
measures the specifically *gravitational* attraction that
two particles exert on each other. *A priori* it is not
evident that these two numbers are the same. However,
it had been known since the time that Newton first
formulated his theory of universal gravitation that they
must be very nearly the same, if not identical. The
primitive observation that suggests this is that, in the
absence of air resistance, a condition possible in experiments done in vacuua, all objects fall with the same
acceleration in the gravitational field of the earth. (This
acceleration is the famous "g" of rocket flight and is
about 32 feet per second per second.) This is true
whatever the "mass" of the object is. In the gravitational
equation the "gravitational mass" of the object cancels
out against its "inertial mass," and the only mass that
remains is the "gravitational mass" of the earth.[3] This
curious feature of gravitation had been recognized
and studied prior to Einstein—the Hungarian nobleman
Baron von Roland Eötvös made a career, early in the
century, of measuring the equality of the inertial and
gravitational mass of high accuracy, and experiments

[3] To put the matter specifically, Newton's law states that
for any force

$$F = ma$$

where m is the "inertial mass," whereas Newton's law of
universal gravitation states that for gravity forces

$$F = \frac{G m M}{r^2}$$

where the masses m and M are "gravitational" masses and
G is the universal "constant of Gravitation." On the surface of the earth we can replace r by the radius of the earth
and M by the mass of the earth. Hence if inertial and gravitational masses are the same we conclude that all objects
near the surface of the earth fall with an acceleration g.

done in the last few years by R. H. Dicke have demonstrated that the two masses are equal to about one part in *100 billion*—but no one, apparently, had thought of this as something worthy of special emphasis. To Einstein, however, this equality smacked not of an accident but rather of a *conspiracy*. In some historical notes that he wrote later he said, "This law, which may be formulated as the law of the equality of inertial and gravitational mass, was now brought home to me in all its significance. I was in the highest degree amazed at its persistence and guessed that in it must lie the key to a deeper understanding of inertia and gravitation. I had no serious doubts about its strict validity even without knowing the results of the admirable experiments of Eötvös which—if my memory is right—I only came to know later."[4]

Einstein's 1911 paper is an attempt to set this observation into a more general framework. It was not a fully successful attempt, however, because at this stage, Einstein had still not abandoned Newton's theory of gravitation. Rather he was adjoining some new extra principles to the old Newtonian theory in an amalgam that did not stick together. However, some of the concepts he introduced then have survived, and the paper is interesting for what it reveals about the evolution of Einstein's thought. One can sense Einstein's struggle in a way that one cannot in the 1916 paper, which stands with an austere and finished perfection. The main new idea in the 1911 paper is what has come to be known as the "principle of equivalence." This is a kind of relativity principle but of a new character. It states that the effects of a uniform constant acceleration on an observer or his measuring instruments are indistinguishable from—that is, equivalent to—the observer's being at rest but acted on by a uniform field of gravitation.

[4] *Essays in Science*, p. 80.

In anticipation of what is to come, and to avoid mis-understandings, it should be emphasized here that this statement of the principle of equivalence is not compatible with the special theory of relativity. The reason is that according to Newton's law, an object placed in a uniform gravitational field will accelerate to arbitrary speeds and will eventually go faster than light. In reality there is in nature no such thing as a truly uniform gravitational field, although for many practical purposes there are regions of space where the gravitational field is very nearly uniform. The correct interpretation of the principle is that if we take a very small region around a given point, the field will vary little in this region, and by taking the region smaller and smaller, we can reduce the variation as much as we choose. In such an infinitesimal region we can replace the local gravitational field by a uniformly accelerating coordinate system. Hence a conflict with the special relativity theory is avoided, since the uniform acceleration takes place over only an infinitesimal region and the particle velocity cannot exceed that of light.

To see what this principle of equivalence means we can imagine ourselves in what has come to be known as an "Einstein elevator." This is a closed box sitting in space somewhere which can be tugged, say, "up," by someone outside pulling on a rope, attached to the roof, with a constant force. The occupants of the elevator will feel themselves pressed "down" toward the floor, and the principle of equivalence asserts that this force downward is identical to that which can be produced by a suitably constructed uniform gravitational field acting "downward" on a stationary elevator. The people inside the elevator will not be able to tell which situation they are experiencing. This law implies the equivalence of gravitational and inertial mass, since in effect it equates a gravitational and an inertial force.

Einstein used the principle of equivalence to draw

other consequences that are extremely interesting. Here we wish to focus on only one; namely, the bending of light by the force of gravity.

Let us imagine the following situation: an elevator is attached to its rope and being pulled upward with a constant force and hence a uniform acceleration. We are stationed outside the elevator in the "rest frame" with respect to which the elevator is accelerating. We now fire a beam of light from our rest frame in such a way that the light enters the elevator—by a small window, if you will—on a trajectory that is initially parallel to the elevator floor. What we will observe happening is the floor of the elevator accelerating upward toward the light beam. To us in our rest frame the light ray follows a straight line, while to the people in the elevator the light beam will appear to have been bent down toward the floor in an arc. But if they do not "know" that they are being pulled up, they can, according to the principle of equivalence, conclude that there is a uniform gravitational field in their region of space which is bending the light downward in a curved path. At first look, this seems very strange, because from a Newtonian point of view only objects with mass are acted on by gravity. Hence if a contradiction with the principle of equivalence is to be avoided, it must be that beams of light propagate in a gravitational field as if they had gravitational mass. But this, Einstein pointed out, is to be expected from the mass-energy relationship, $E = mc^2$, of the special theory of relativity. Light beams certainly transport energy—light from the sun, after all, heats the earth. Hence, Einstein argued, if the energy content of the light is E, then its equivalent gravitational mass must be E/c^2.

It should be emphasized that the principle of equivalence alone does not yield a unique prescription for the way light is bent by the gravitational field. Rather, it only suggests that light should be affected by gravity. Having realized this, Einstein, in the 1911 paper, used the Newtonian gravitational force law to compute the trajectories of light rays in a Newtonian gravitational field. When he later saw that Newton's gravitational law had to be modified, he of course had to redo this early computation, using the gravitational dynamics of his 1916 paper. The early calculation is presented first because it forms a useful introduction to the subject and illustrates how scientific ideas evolve a few steps at a time, each step profiting from the errors and partial insights of the previous ones.

At the end of his 1911 paper Einstein noted that his idea that light had an equivalent gravitational mass was, in fact, subject to experimental test. In describing how this was to be accomplished we will first bypass Einstein's argument and give instead an equivalent argument due to the German mathematician and surveyor Johann Georg von Soldner, who had proposed, totally unknown to Einstein, the same effect in 1801![5] This may seem incredible until one realizes that all of the early theories of light involved the idea that light consists of particles of some kind. It was only in the nineteenth century that this notion fell out of favor when experiments apparently "proved" that light behaved in many circumstances as if it were composed of waves of electromagnetic radiation. We now know that light exhibits attributes of both a particle and wavelike character, and that the resolution of this apparent paradox involves, according to most physicists, the quantum theory. Soldner did his work just before the wave nature of light had been demonstrated, and he carried out in detail the response to the first "Query" in Newton's

[5] Soldner's paper was reprinted in the *Annalen der Physik*, 65 (1921), 593, accompanied by a long introduction by Philipp Lenard. Lenard, a well-known experimental physicist and an early and enthusiastic Nazi, said that his motive in having Soldner's paper reprinted was to give support to his claim that Einstein's work had been previously invented by "Aryans."

Opticks where the particle view of light is set forth. "Do not Bodies act upon Light at a distance, and by their action bend its rays, and is not this action (*caeteris paribus*) strongest at the least distance?" By "least distance" Newton meant the distance that is closest to the center of gravity of the gravitating mass which is bending the light. Soldner computed the trajectory of a light particle emitted by a distant star as it passed close to the edge of the sun. He did not have to know the mass of the light particle, since he simply assumed that its inertial mass, whatever that was, canceled out the equivalent gravitational mass of the light particle, so that the only mass that remained in the problem was the mass of the sun, which he knew.

Soldner found that bending the trajectory of the light particle gives rise to an observable effect in the following way. Suppose first that the earth is at some position on its orbit around the sun so that the star in question does not appear in the neighborhood of the sun. At this instant we determine, with telescopes and so on, what we may call the "true" position of the star. Now imagine that the earth has moved around so that light from the star, which has not moved, must pass close to the edge of the sun to reach the earth. According to Soldner's computation the path of such a light ray is bent, as it happens, concave toward the sun. The trajectory is pulled around the sun in an arc, as a space ship would be. This means, to exaggerate the effect for clarity, that we must shift the telescope so that it points toward the edge of the sun, although the "true" position of the star is, again to exaggerate, behind the sun. Shifting the telescope will make it appear to us as if the star has shifted its position—*away* from the sun—by a small angle. Our line of sight passes along the tangent to the edge of the sun and the star will appear along the prolongation of this line of sight—hence away from the sun. In fact, *every* distant star will appear shifted by the *same* angle when its light passes close to the sun, since this angle, according to Soldner's calculation—and Einstein's as well—depends only on properties of the sun; i.e., its mass, and not on which star has emitted it. The angle predicted by this computation is incredibly small: 83 seconds of arc. Of course, under most circumstances the sun shines so intensely that one cannot carry out this experiment in practice, since pointing a telescope directly at the sun, which means during the day, is hardly going to show up any stars. But, Einstein suggested in 1911, "As the fixed stars in the parts of the sky near the sun are visible during total eclipses of the sun, this consequence of the theory may be compared with experiment." And he adds, "With the planet Jupiter the displacement to be expected [for a star in the neighborhood of Jupiter] reaches about 1/100 of the amount given. [Jupiter has a smaller mass and hence there is a smaller effect.]" He concludes, "It would be a most desirable thing if astronomers would take up the question here raised. For apart from any theory there is the question whether it is possible with the equipment at present available to detect an influence of gravitational fields on the propagation of light."[6]

In fact in 1914, just before the outbreak of the war, an expedition of German astronomers left for Russia, where a total eclipse was predicted to be observable. But before they could make any measurements they were interred as prisoners of war. (After a few weeks they were released, minus their equipment, in exchange for some Russian officers.) If they had carried out their measurements they would have found a result that certainly would have interested Einstein but probably not, at this point in the evolution of his ideas, surprised him very much. Namely, they would have found the effect, but an effect *twice as big as* the one he predicted in 1911— indicating that at least some of the assumptions of the

[6] *The Principle of Relativity*, p. 99.

1911 paper were wrong. As it happened, Einstein soon became convinced of this anyway.

Before we go into the details it is worth while to make a few general comments. As mentioned earlier, it was at this point in Einstein's work that he made almost a "quantum leap" in levels of mathematical abstraction. Along with this he appears to have made a similar philosophical leap, the success of which influenced his outlook on science forever afterward. The early Einstein papers seem to be rooted in a sort of clairvoyant view of the meaning of physical phenomena. There is an overriding sense of being close to the physical phenomena even when they are being described in new and apparently revolutionary terms. On the other hand, in the leap that led him to general relativity, the connection with the phenomena is exceedingly indirect. He is guided not by experiments—the experiments came several years after the theory had been published—but by philosophical or epistemological principles. A critic might say that he was influenced by philosophical, or even metaphysical, *prejudices*. But the fact that these could lead, in his hands, to a physical theory of incredible power, convinced him of the power inherent in the human mind to comprehend the physical universe. In June 1933 he delivered the Herbert Spencer lecture at Oxford in which he attempted to analyze what he called "the Method of Theoretical Physics." The title is somewhat misleading because what he really describes is *his* method of doing theoretical physics, which was, by this time, almost completely distinct from that of any of his contemporaries. In fact in some bizarre sense his "method" has more in common with the philosophical attitudes of Plato, with the Platonic emphasis on perfect shapes and forms, than with any physicist one can think of since and including Newton.

Our experience hitherto justified us in believing that nature is the realization of the simplest conceivable

mathematical ideas. I am convinced that we can discover, by means of purely mathematical constructions,
those concepts and those lawful connections between
them which furnish the key to the understanding of
natural phenomena. Experience may suggest the
appropriate mathematical concepts, but they most
certainly cannot be deduced from it. Experience
remains, of course, the sole criterion of physical util-
ity of a mathematical construction. But the creative
principle resides in mathematics. In a certain sense,
therefore, I hold it true that pure thought can grasp
reality, as the ancients dreamed.[7]

The guiding philosophical principle that underlies
the general theory of relativity is Einstein's conviction
of the meaninglessness of absolute empty space as a
physical entity. As he sometimes put it, "Space is not a
thing." Space and time are only given meaning in terms
of meter sticks and clocks. This, as we have seen, played
a key role in the formulation of the special theory of
relativity, but it returns in the general theory of relativity in a more sophisticated guise. The "special theory" is "special" because it deals with one sort of motion
—uniform motion with constant velocity in a straight
line. Newton himself believed that such motions could
not be distinguished, from the point of view of the laws
of physics as he knew them, from a state of rest. But in
his view accelerations were something different altogether. When we are accelerated we *feel* it—we are
pushed or pulled—and hence it would appear to be
reasonable to claim that we can measure accelerations,
even in "empty" space, in some absolute sense.

Newton's most frequently discussed examples had to
do with rotations. Imagine, for example, that we have
two weights connected by a string and that we set these
weights rotating around some central point. It is then a
fact of common experience that each weight will pull

[7] *Essays in Science* (where the lecture is reprinted in full),
p. 17.

against the string. It would appear and did appear to
Newton as if we could use the amount of tension in the
string as a measure of the *absolute acceleration* in empty
space of these rotating weights. All of this may sound
reasonable until one begins to analyze the forces that,
from the Newtonian point of view, must act on the
bodies in question to produce the observed rotation. A
particularly clear example of what is involved is the
stationary orbit of the synchronous satellites used to re-
lay radio signals. As is well known, to produce such an
orbit one arranges things, for, say, a satellite fired parallel to the Earth's equator, so that the rate of rotation of
the satellite, as viewed from the North Pole, exactly
matches the rate at which a point on the equator rotates
about the pole. Hence, as seen from the equator, the
satellite hovers over a fixed position on the equator.
Someone viewing the satellite from the equator will
conclude, from the Newtonian point of view, that since
the satellite exhibits no acceleration—it simply rests
stationary overhead—there must be, by Newton's law,
no net force acting on it. However, we know that the
force of gravity is pulling the satellite *down*, hence, a
Newtonian would be forced to argue, there must be an
opposing force, which just balances the force of gravitation, which is pushing the satellite *up*. In elementary
physics books this force is even given a name. It is called
the "centrifugal force." Although beginning physics students learn to use the equations that govern this "force,"
they often feel a certain ill ease about it, and with good
reason.

Unlike the other forces of physics the centrifugal
force does not appear to be associated with the influence on a given object of other neighboring objects. The
force of gravitation is produced by the neighboring
presence of material bodies; electrical forces are produced by the presence of electrically charged objects;
and so on. But the centrifugal force appears to be produced by the relationship of an object to *empty space*.

In his 1916 paper on general relativity Einstein gave a graphic illustration of how peculiar this situation is. He imagined a universe that consisted of two identical spheres made of some deformable plastic material. Now according to Newtonian physics we can conceive of a situation in which one sphere is deformed into an ellipsoid while the other sphere remains perfectly spherical even though both spheres are totally removed from the influence of any external objects and even though each sphere produces the same influence on the other one: this is the situation in which one sphere is spun so that it rotates on its axis while the other remains at rest. The rotating sphere will experience "centrifugal forces" which will cause it to bulge—or, at least, this is what the Newtonian would argue. But if one asks what causes the centrifugal force, the only answer that a Newtonian can give is the rotation of the sphere with respect to empty, or absolute, space. In fact in many elementary physics books these centrifugal forces are often referred to as "fictitious forces," since they disappear once the system that is rotating has been brought to rest. One may ask why it is necessary to introduce such forces in the first place? From the Newtonian point of view, the answer is very clear—without them Newton's law that relates force to acceleration is false. This is clearly illustrated by the case of the synchronous satellite mentioned above. The only "real" force acting on the satellite is the force of gravity, and hence, if the centrifugal force had not been introduced, Newton's law would have predicted that the satellite should fall to earth, which contradicts experience.

The first person to criticize this Newtonian point of view was, interestingly enough, Newton's contemporary the Irish philosopher Bishop Berkeley. Twenty years after the publication of the *Principia* Berkeley argued, in astonishingly modern terms, that attributing forces to the influence of absolute space was nonsense, and he went so far as to hint that perhaps what was involved was some influence of distant rotating stars. His contemporaries thought *this* was nonsense, and there was, in any case, no way to fit it into the context of Newtonian gravitation. The matter was more or less dropped until the end of the nineteenth century. The real forerunner of the modern point of view and the man whose influence on the young Einstein was really decisive was the Austrian physicist, philosopher, physiologist, and general polymath Ernst Mach. Mach is known to the nonphysicist, if at all, by the so-called Mach numbers: Mach 1, for example, is the speed of sound in air at a standard temperature and pressure, and is relevant in the dynamics of air flow over a moving object—an airplane wing—when that object moves at the speed of sound. (Mach did significant experimental work on this question and hence the Mach numbers.) However, to his contemporaries Mach was an influential intellectual giant. In 1882, William James, who heard him lecture in Prague, where Mach became the first rector of the German University to which Einstein went in 1910, wrote, "I don't think anyone ever gave me so strong an impression of pure intellectual genius. He apparently has read everything and thought about everything, and has an absolute simplicity of manner and winningness of smile when his face lights up, that are charming."[8]

Mach had a devastating critical insight into the foundations of physics, and he took it upon himself to rid physics of what he felt was "metaphyics," by which he meant any element that could not be connected directly to sense experience. His judgment was not always infallible—he rejected until near the end of his life (in 1916) the concept of atoms—and Einstein, after the success of general relativity, rejected "Machism" in its simplest form. In 1917 Einstein wrote to his friend from

[8] Cited in Gerald Holton, "Mach, Einstein and the Search for Reality," p. 664. The article gives a full discussion of the relationship between Einstein and Mach.

the patent office, Besso, who had first introduced him to Mach's writings when they were students in Zurich,

> and mentioned a manuscript which Friedrich Adler had sent him. Einstein commented: "He rides Mach's poor horse to exhaustion." To this Besso—the loyal Machist—responds on 5 May 1917: "As to Mach's little horse, we should not insult it; did it not make possible the infernal journey through the relativities? And who knows—in the case of the nasty quanta, it may also carry Don Quixote de la Einstina through it all!" . . . Einstein's answer of 13 May 1917 is revealing: "I do not inveigh against Mach's little horse; but you know what I think about it. It cannot give birth to anything living, it can only exterminate harmful vermin."[9]

The book that Besso had given Einstein in 1897 was Mach's *Science of Mechanics,* one of the most important critical works in the development of modern physics, and the "vermin" that it exterminated was Newton's conception of absolute space. Every page of the *Mechanics* brims over with a sort of polemical fury and with what Einstein referred to as Mach's "incorruptible skepticism and independence."[10] Here is a typical passage:

> It is scarcely necessary to remark that in the reflections here presented, Newton has again acted contrary to his expressed intention only to investigate *actual facts.*
> No one is competent to predicate things about absolute space and absolute motion; they are pure things of thought, pure mental constructs, that cannot be produced in experience. All our principles of mechanics are, as we have shown in detail, experimental knowledge concerning the relative positions

[9] *Ibid.*, p. 657.
[10] Paul Arthur Schilp, ed., *Albert Einstein, Philosopher-Scientist,* p. 21.

and motions of bodies. Even in the provinces in which they are now recognized as valid, they could not, and were not, admitted without previously being subjected to experimental tests. No one is warranted in extending these principles beyond the boundaries of experience. In fact, such an extension is meaningless, as no one possesses the requisite knowledge to make use of it.[11]

Of course it is one thing to be able to criticize scientific ideas, and it is another to be able to create new ones, and it is not always the case that these two faculties are present in the same individual. Perhaps if Mach had been younger when the *Science of Mechanics* was published, which was in 1883 when Mach was forty-five, he might have discovered the general theory of relativity. In fact he did not, but he did make an important restatement of Bishop Berkeley's idea that distant matter in the universe may influence the behavior of accelerated objects; i.e., may give a physical explanation of "centrifugal force." His formulation has now become known as Mach's Principle. In particular Mach dismissed Newton's examples of isolated bodies undergoing rotation, because he said that in the actual universe that were no such things as isolated bodies since one could not in reality turn off the effects of the distant stars. One had no experimental evidence as to what a body—one of Einstein's spheres—would do if it rotated in an empty universe, since, as far as we are concerned, there is no such thing as an empty universe. As Mach put it:

> For me only relative motions exist. . . . When a body rotates relatively to the fixed stars, centrifugal forces are produced; when it rotates relative to some different body and not relative to the fixed stars, no centrifugal forces are produced. I have no objection to just calling the first rotation so long as it be remembered

[11] Quoted in Arnold Koslow, *The Changeless Order*, p. 143.

that nothing is meant except relative rotation with respect to the fixed stars.[12]

Einstein realized from the beginning—as early as 1908, he tells us in his notes[13]—that the special theory of relativity could not be a satisfactory basis for a theory of gravitation that would include the principle of equivalence and Mach's principle. The reason is that the special theory is based on a geometric view of space and time that is not broad enough to include the effects of acceleration and, hence, gravity. This basic point can be made with a simple example again often used by Einstein. We may imagine a flat circular disc that has been set into rotation around its central point—a spinning wheel. According to classical physics the geometry of this disc remains Euclidean even when the disc is rotating. By Euclidean, in this case, one means that if one measures the circumference of the disc and divides it by the diameter of the disc the answer is always the celebrated number known as "pi." (To several decimal places pi is given by 3.1415927. . . .) This is true of all circles in a Euclidean universe. However, the relativity theory teaches us that a line segment on the perimeter of a rotating circle will undergo a Lorentz contraction as measured by an observer at the fixed central point about which the perimeter is rotating. On the other hand the diameter of the circle is not Lorentz-contracted, since it is at right angles to the direction of the velocity of a point on the perimeter. (In an actual circle drawn with a pencil the diameter would have some width and this width would be Lorentz-contracted, but we are imagining here an ideal circle with an ideal diameter, which is a line with no width.) Thus the measured ratio of circumference to diameter will no longer be pi accord-

[12] From Mach's *History and Root of the Principle of the Conservation of Energy*, cited in D. W. Sciama, *The Physical Foundations of General Relativity*, p. 18.
[13] "Notes on the Origin of the General Theory of Relativity," *Essays in Science*, p. 78.

ing to the relativity theory, and hence the geometry will no longer be Euclidean once accelerations are considered.

Similar remarks apply to clocks. As viewed by an observer at the central point, clocks on the perimeter of the disc, and indeed at any intermediary point, will be "slow" as compared to the central clock. Thus on the disc it will not be possible to set up any over-all space-time coordinate system in the sense of the special relativity theory. One cannot make such a system by laying out rulers and clocks, since these contract, in the case of the rulers, or slow down, in the case of the clocks, in varying degrees from point to point on the disc as they are viewed by the central observers. This dilemma was the great obstacle that prevented Einstein from making an immediate generalization of the special theory to accommodate accelerated motions.

> I soon saw that bringing in non-linear transformations [generalizing the Lorentz transformations to include accelerations] as the principle of equivalence demanded, was inevitably fatal to the simple physical interpretation of the co-ordinates—i.e., that it could no longer be required that differentials of co-ordinates should signify direct results of measurement with ideal scales or clocks. I was much bothered by this piece of knowledge, for it took me a long time to see what co-ordinates in general really meant in physics. I did not find the way out of this dilemma till 1912.[14]

The "way out of the dilemma" is not easy to describe for readers without mathematical training. However, it has modified our views of space and time and the cosmos in such significant ways that it is worth trying to convey at least the essence of it. We shall here give a somewhat abbreviated didactic overview and come back to it later in more detail. The essence of Einstein's

[14] *Ibid.*, p. 81.

new view is that there is a hitherto unsuspected connection between the geometry of space-time and gravitation. We have seen a hint of this in our discussion of the bending of light as Einstein presented it in his 1911 paper. After all, the whole of geometry depends on the behavior of "straight" lines, and ultimately we determine the straightness of lines by using light and the propagation of light. If light beams do not obey a Euclidean geometry in the presence of gravitation then our notions of physical geometry must be modified. In practice these effects are small because, as we have pointed out, gravitation is under most circumstances the weakest force we know of. However, under certain circumstances these small effects become measurable—the bending of light by the sun, for example. However, the 1911 paper is unsatisfactory, since it takes Newtonian gravitation as a separate given entity—something known and fixed—that acts on light. And, as we have argued above, Newton's law of universal gravitation is unsatisfactory because on one hand it gives no account of the principle of equivalence and on the other it depends on the action of absolute space.

In Einstein's new approach the equations of Newtonian gravitation are replaced by a new set of equations that have the property that they retain their form in *all possible coordinate systems*, both accelerated and uniformly moving. Einstein arrived at these equations by postulating that they should be the simplest mathematical expressions consistent with this general invariance. Hence the equations are essentially forced on one, once one makes the assumption that absolute space can play no role in physics. These equations determine the geometry of space-time. Where there is no gravity— i.e., very far away from massive objects—this geometry is the pseudo-Euclidean geometry of the Minkowski world of special relativity. This geometry is called "flat" because light propagates in straight lines which behave according to the axioms of Euclid. But in the presence of gravitation the geometry of space-time is altered, which is sometimes expressed by the statement that space-time becomes "warped" or "curved"; this means simply that figures constructed out of light rays do not satisfy Euclidean geometry if the light rays propagate in the presence of gravitating matter. Nonetheless it is meaningful to say that light rays still propagate in "geodesics," which are the curves that play the role of straight lines in the new geometry.[15] For example, the great circles drawn for longitude and latitude on the surface of a globe are the "geodesics" of this spherical geometry. In certain simple cases Einstein's equations can be solved approximately. In the approximation in which gravitational effects are weak they reduce to Newton's laws, although with a new geometric interpretation. Hence the gross structure of planetary motion is still given nearly correctly by the Newtonian laws. How-

---

[15] At this point it is perhaps worth while to make a significant technical observation. In the 1916 paper Einstein had to make, essentially, a *postulate* that material objects, as well as light, followed geodesic paths in a gravitational field. However, in 1938, in collaboration with Infeld and Banesh Hoffmann, he was able to show that in a sense this postulate was not needed. In this work a material object is abstractly characterized as a point in space-time where the gravitational field becomes infinitely strong—a "singularity." He was able to show that solutions to the field equations could be obtained of such a character that they contain singular points only along geodesics, which means that "particles" necessarily follow these paths. This work was perhaps the last done by Einstein on which all physicists could agree to its significance and importance. And his success here, as much as anything, appears to have convinced him that he was on the right track toward an ultimate description of matter. He never was able to carry out the rest of the program, which would have to include the other forces besides gravitation, and most physicists believe that it cannot be carried out without including elements from the quantum theory, which Einstein was not prepared to do. Hence he worked along this path in near isolation for the rest of his life. (I am grateful to Freeman Dyson for calling my attention to this aspect of Einstein's later work.)

ever, for a light ray moving near the sun, there are measurable corrections to the Newtonian predictions. The space is sufficiently "curved" by the gravitation of the sun that the light geodesic is different from the Newtonian prediction. In fact the new theory predicts that there should be an apparent shift in the position of stars whose light passes close to the sun's surface of 1.74 seconds of arc, which is *twice* as much as the Newtonian theory used in Einstein's 1911 paper predicted.[16]

This prediction was published by Einstein in his 1916 paper, written during the height of the First World War. Not surprisingly, it was not tested until 1919, when the war had ended.

[16] It is again important to note that the difference between this prediction and that of the 1911 paper is not simply the numerical factor, but rather the entire theoretical basis. In the 1916 theory part of the light-bending is given by a modification of the geometry of space and part by the fact that clocks run slow in a gravitational field—a point to which we return shortly. The arguments of the 1911 paper lead to a correct prediction for the slowing of the clocks, but not for the change in the spatial geometry. Hence, one can say that experiments that confirm the slowing of the clocks really confirm only the principle of equivalence, while the light-bending confirms the idea that the geometry of both space and time—space-time—is determined by gravity. This is why so much effort has gone into repeating and refining these measurements.

The First World War and the Bending of Light

xi

In 1914, shortly after Einstein moved from Zurich to Berlin, the war began.

The years of the First World War were both very difficult and very happy ones for him. By 1914, shortly after he came to Berlin, he and his wife separated and she returned to Zurich with their two young sons. But during the war Einstein rediscovered parts of his family who were living in Berlin—or rather, they rediscovered *him*. He had long been thought of as an irresponsible Bohemian by his rather wealthy and staid middle-class Berlin relatives, but now that he was a member of the Royal Prussian Academy, they were pleased to have him in the family. This was fortunate for Einstein in at least two ways. Einstein was frequently in poor health—no doubt due in part to malnutrition, since as a bachelor once again his diet was left to the vagaries of the Berlin restaurants in war-

time. At his cousin Rudolph's house he was well fed, and
he also had the companionship of Rudolph's daughter
Elsa, recently widowed with two daughters, a "woman of
friendly, maternal temperament, fond of amusing conversation and interested in creating a pleasant home."[1]
When Einstein married her, in 1919, there was some
criticism in Berlin academic circles that she was not
sufficiently "intellectual" for him. "But," as Philipp
Frank noted, "if Einstein had followed this criticism,
what woman could he have married? . . . The married
life of a great man has always been a difficult problem,
no matter how he or his wife is constituted. Nietzsche
once said: 'A married philosopher, is, to put it bluntly,
a ridiculous figure.' "[2] Be this as it may, there is every
evidence that Einstein's second marriage was a happy
one and that his cousin had a great talent for giving
him the kind of serene home life that he needed for
doing his work.

From his earliest childhood Einstein had deeply
rooted pacifist feelings. He simply hated and despised
anything that represented war and soliders—no weaker
words convey his feelings. Hence when the First World
War broke out he refused to have anything to do with
the celebrated *Manifesto to the Civilised World* which
ended with the sentiment that German culture and
German militarism had to be accepted together, and
which was signed by ninety-three prominent German
artists, scientists, and writers (including Planck). In
fact, he tried to involve himself with pacifists in other
countries, including Romaine Rolland, author of *Jean-
Cristophe*. Rolland, who had fallen into disfavor in
France, was living near Geneva, and Einstein came to
visit him in September 1915 after a visit to his children
in Zurich. Rolland recorded some impressions of this
visit in his diary:

[1] Philipp Frank, *Einstein: His Life and Times*, pp. 123–24.
[2] *Ibid.*, pp. 125–26.

Einstein is still a young man [he was then thirty-six], not very tall, with a wide and long face, and a great mane of crisp, frizzled and very black hair, sprinkled with gray and rising high from a lofty brow. His nose is fleshy and prominent, his mouth small, his lips full, his cheeks plump, his chin rounded. He wears a small cropped mustache. He speaks French rather haltingly, interspersing it with German. He is very much alive and fond of laughter. He cannot help giving an amusing twist to the most serious thoughts.

Einstein is incredibly outspoken in his opinion about Germany, where he lives and which is his second fatherland (or his first). No other German acts and speaks with a similar degree of freedom. Another man might have suffered from a sense of isolation during that terrible last year, but not he. He laughs. He has found it possible, during the war, to write his most important scientific work. I ask him whether he voices his ideas to his German friends and whether he discusses them with them. He says no. He limits himself to putting questions to them, in the Socratic manner, in order to challenge their complacency. People don't like that very much, he adds.[3]

Einstein himself wrote to his friend the physicist Paul Ehrenfest:

Europe, in her insanity, has started something unbelievable. In such times one realizes to what a sad species of animal one belongs. I quietly pursue my peaceful studies and contemplations and feel only pity and disgust. My dear astronomer Freundlich [the leader of the ill-fated 1914 expedition] will become a prisoner of war in Russia instead of being able there to observe the eclipse of the sun. I am worried about him.[4]

Einstein continued his "peaceful studies" throughout the war—he wrote some thirty papers between 1915 and

[3] Otto Nathan and Heinz Norden, eds., *Einstein on Peace*, p. 14.
[4] *Ibid.*, pp. 14–15.

1918—but, as we have mentioned, there was no possibility of testing the general relativity theory until the war ended. No doubt an attempt would have been made to test Einstein's theory of light deflection by the sun sooner or later. But the fact that two expeditions were launched by the British in 1919, so soon after the war, was certainly due to Sir Arthur Eddington's interest in and understanding of the revolutionary implications of general relativity. (Moreover, Eddington was a Quaker, which made him especially conscious of the opportunity that such an expedition would afford to repair the scientific alliances broken by the war.) Eddington had been deeply attracted to the theory from the beginning. In fact he was so convinced of its truth that when Sir Frank Dyson, the Astronomer Royal, was asked by one of the members of Eddington's group what would happen if the observations gave *twice* the Einstein prediction, Dyson replied, "Then Eddington would go mad, and you would have to come home alone."[5]

There is no way to improve on Eddington's own description of what happened in 1919, as he recounted it in his classic popular book on relativity, *Space, Time and Gravitation*:

> In a superstitious age a natural philosopher wishing to perform an important experiment would consult an astrologer to ascertain an auspicious moment for the trial. With better reason, an astronomer of today consulting the stars would announce that the most favorable day of the year for weighing light is May 29. The reason is that the sun in its annual journey round the ecliptic goes through fields of stars of varying richness, but on May 29 it is in the midst of a quite exceptional patch of bright stars—part of the Hyades—by far the best star field encountered. Now

[5] Eugene Guth, "Contribution to the History of Einstein's Geometry as a Branch of Physics," *Proceedings of the Relativity Conference in the Midwest 1969* (New York, 1970), p. 186.

if this problem had been put forward at some other period of history, it might have been necessary to wait some thousands of years for a total eclipse to happen on the lucky date. But by a strange good fortune an eclipse did happen on May 29, 1919.

In March 1917 the Astronomer Royal alerted the British astronomers to this chance, and preparations for two small expeditions were begun. One was to go to Sobral, in northern Brazil, and the other, Eddington's, to the Isle of Principe in the Gulf of Guinea, off West Africa—both being located on the path of totality of the eclipse. Eddington and another astronomer arrived in Principe in the spring of 1919 with a large telescope and a good deal of photographic material, and they spent over a month there.

On the day of the eclipse the weather was unfavorable. When totality began the dark disc of the moon surrounded by the corona was visible through cloud, much as the moon often appears through cloud on a night when no stars can be seen. There was nothing for it but to carry out the arranged programme and hope for the best. One observer was kept occupied changing the plates in rapid succession, whilst the other gave the exposures of the required length with a screen held in front of the object-glass to avoid shaking the telescope in any way.

For in and out, above, about, below
'Tis nothing but a Magic *Shadow*-show
Played in a Box whose candle is the Sun
Round which we Phantom Figures come and go.

Our shadow-box takes up all our attention. There is a marvelous spectacle above, and, as the photographs afterwards revealed, a wonderful prominence-flame is poised a hundred thousand miles above the surface of the sun. We have no time to snatch a glance at it. We are conscious only of the weird half-light of the landscape and the hush of nature, broken by the calls of the observers, and beat of the metronome ticking out the 302 seconds of totality. . . .

[We obtained sixteen photographs of which only] one was found showing fairly good images of five stars, which were suitable for a determination. This was measured on the spot a few days after the eclipse in a micrometric measuring-machine. The problem was to determine how the apparent positions of the stars, affected by the sun's gravitational field, compared with the normal positions on a photograph taken when the sun was out of the way. Normal photographs for comparison had been taken with the same telescope in England in January. The eclipse photograph and a comparison photograph were placed film to film in the measuring machine so that corresponding images fell close together, and the small distances were measured in two rectangular directions. From these the relative displacements of the stars could be ascertained. . . .

The results from this plate gave a definite displacement, in good accordance with Einstein's theory and disagreeing with the Newtonian prediction. Although the material was very meagre compared with what had been hoped for, the writer (who it must be admitted was not altogether unbiased) believed it convincing.[6]

The Principe experiment gave a displacement of 1.61 seconds of arc with an experimental error margin of 30 seconds, while the Sobral data gave 1.98 seconds with an error of 12 seconds; both results were in reasonable agreement with Einstein's prediction of 1.74 seconds. The evidence was thought to be sufficiently solid so that it was presented at a joint session of the Royal Society and the Royal Astronomical Society in London in November 1919. Considering what was at stake and the connection of Newton to the beginnings of the Royal Society, the atmosphere of this meeting must have been

[6] Sir Arthur Eddington, *Space, Time, and Gravitation*, pp. 113–22 *passim*.

extraordinary. It is captured in a description by Alfred North Whitehead, who was there:

> The whole atmosphere of tense interest was exactly that of the Greek drama. We were the chorus commenting on the decree of destiny as disclosed in the development of a supreme incident. There was dramatic quality in the very staging—the traditional ceremonial, and in the background the picture of Newton to remind us that the greatest of scientific generalizations was now, after more than two centuries, to receive its first modification. Nor was the personal interest wanting; a great adventure in thought had at length come safe to shore.[7]

There are two glimpses of Einstein's own reaction to the first experimental confirmation of his theory. The first is a reminiscence by Ilse Rosenthal-Schneider, a student of Einstein in 1919 who was with him soon after he had received word about the eclipse expedition.

> He suddenly interrupted the discussion . . . reached for a telegram that was lying on the windowsill, and handed it to me with the words, "Here, this will perhaps interest you." It was Eddington's cable with the results of the eclipse expedition. When I was giving expression to my joy that the results coincided with his calculations, he said quite unmoved, "But I knew that the theory is correct"; and when I asked what if there had been no confirmation of his prediction he countered, "Da könnt' mir halt der liebe Gott leid tun, die Theorie stimmt doch." . . . "Then I would have been sorry for the dear Lord—the theory is correct."[8]

And on September 27, 1919, soon after he had received a cable from Lorentz—"Eddington found star displacement at rim of sun"—who had heard about the re-

[7] Alfred North Whitehead, *Science and the Modern World* (Cambridge, 1933), p. 13.
[8] Quoted in Gerald Holton, "Mach, Einstein and the Search for Reality," p. 653.

sults of Eddington's analysis of the data before they had become official, Einstein wrote a postcard to his mother (she was then in a sanitorium in Lucerne) that mostly has to do with family matters but begins, "Joyful news today. H. A. Lorentz has telegraphed me that the English expedition has really proved the deflection of light by the sun."

We shall review the more recent evidence for the general theory of relativity, but here it is interesting to present a partial list of the results of the successful eclipse measurements through the year 1952.[9] The first entry gives the sponsoring observatory, the second the location and date of the observation, and the third the result, with the quoted error (in seconds of arc).

Greenwich—Australia, September 21, 1922	1.77	.40
Potsdam—Sumatra, September 21, 1929	1.82	.20
Sternberg—U.S.S.R., June 19,1936	2.73	.31
Sendai—Japan, June 19, 1947	2.13	1.15
Yerkes—Brazil, May 20, 1952	2.01	0.27
Yerkes—Sudan, February 25, 1952	1.70	0.10

The fact that these numbers have a rather wide "scatter" indicates how difficult precision observational astronomy really is. If this were the only evidence for a theory this fundamental, one would be left with a certain sense of ill ease. But there is, as we shall see, a great deal of other independent evidence involving entirely different kinds of phenomena. And then there is the extraordinary beauty and harmony of the theory as a whole. It is as if, as Einstein might have put it, the Old One has allowed a brief lifting of the clouds of mystery

[9] The list is taken from D. W. Sciama, *The Physical Foundations of General Relativity*, p. 71. Professor Sciama, now of Oxford, participated in a 1954 expedition that ran into bad weather and could not take any pictures. This is only part of Sciama's list but enough to give a flavor of the work and of its international character.

that surround the ultimate secrets of the workings of the world. And we, through Einstein's genius, have been given a new vision.

For Einstein, himself, this was the end of his life as a private person. Almost upon the day of the public announcement of the bending of starlight by the sun he became a public figure—a sensation—and a symbol. This change also coincided with the first serious anti-Semitic outbreaks in Germany (which was then in a near state of civil war), some of which were, even then, directed against Einstein and "Jewish physics." It was perhaps with these things in mind that when *The Times* in London asked him for an explanation of the theory of relativity for the general reader he wrote an article on November 28, 1919, to which he added the postscript:

> Some of the statements in your paper concerning my life and person owe their origin to the lively imagination of the writer. Here is yet another application of the principle of relativity for the delectation of the reader: —Today I am described in Germany as a "German savant," and in England as a "Swiss Jew." Should it ever be my fate to be represented as a *bête noire*, I should, on the contrary, become a "Swiss Jew" for the Germans and a "German savant" for the English.[10]

[10] *Essays on Science*, p. 60.

Geometry and Cosmology

XII

As we have said, the principle result of Einstein's theory of gravitation is that the presence of gravitation, the effect of a gravitating body like a planet or the sun, is to modify the geometry of space and, more generally, of space and time. It is easy to get carried away into fantasy and confusion as to what this means, and so it is important to make the discussion very concrete. In the geometry of Euclid every triangle has three angles whose sum adds up to exactly 180 degrees. This is what is *meant* by a triangle in Euclidean geometry. However, we can draw a physical triangle on the ground with a stick and ask if *this* triangle has an angle sum of 180 degrees. Strictly speaking the answer is *no* because the ground is part of the earth and the earth is a sphere—a globe—and triangles drawn on a globe do not have angle sums of 180 degrees. (The reader may take a globe and draw a triangle with the apex at the North Pole and the base on the equator. He will then find that such a triangle has an angle sum that is *greater* than 180 degrees.) It is a question of physical measurement whether a given triangle constructed in nature is, or is not, Euclidean. The question of the existence on non-Euclidean geometrics was raised by the German mathematician Karl Friedrich Gauss in the late eighteenth and early nineteenth century; Gauss, and afterward Johann Bolyai and Nikolai Lobachevski, constructed perfectly consistent geometries which were non-Euclidean. One could prove all sorts of correct theorems in these geometries, even though triangles do not necessarily have angles that sum up to 180 degrees. This work was further extended by the nineteenth-century German mathematician Bernhard Riemann.

One may summarize the situation as follows: there are three general types of geometry: first there is the familiar Euclidean geometry characterized by 180-degree triangles. This geometry is said to be "flat" because it is the geometry of a flat planar surface. Next there is the geometry studied by Riemann, in which all triangles have angle sums *greater* than 180 degrees. This geometry is called "elliptic," or the geometry of a surface with "positive" curvature, like a sphere. Finally there is the so-called "hyperbolic" geometry of Gauss and his successors, in which all triangles have angle sums *less* than 180 degrees. One can visualize this as the geometry of a surface of "negative" curvature shaped something like a funnel.

In 1900 it occurred to the German astronomer Karl Schwarzschild of Göttingen to try actually to measure the geometry of space by using the light rays from a star as they intersected the earth at two widely separated points of it; orbit around the sun. This makes a huge triangle, and Schwarzschild wanted to know what its angle sum would be and hence what the "curvature" of space was. He did not have any particular theory in mind, apparently, but simply was intrigued by the question. As he wrote:

> One finds oneself here, if one will, in a geometrical fairyland, but the beauty of this fairy tale is that one does not know but what it may be true. We accordingly bespeak the question here of how far we must push back the frontiers of this fairyland; of how small we must choose the curvature of space, how great its radius of curvature.[1]

He performed measurements on various stars but could only conclude that if space is curved the radius of curvature must be very large. (An infinite radius of curvature, by definition, corresponds to flat, i.e., Euclidean, space.) It is somehow fitting that it was Schwarzschild who, in 1916, the year he died, found the first exact solution to Einstein's field equations for a situation in which the gravitational source is taken to be a single fixed mass point. This is a very good approximation for many problems in gravitational physics and the "Schwarzschild solution" is still of fundamental importance in the physics of gravitation. It gives the main features of the gravitational motion of a planet, which is light and mobile compared to the sun, taken in this approximation as a fixed mass point.

By now the reader will, perhaps, have become accustomed to the fact that in the relativity theory modifications of space are usually conjoined with modifications of time: i.e., the behavior of clocks. Hence it may not be altogether surprising that in Einstein's theory of gravitation the behavior of a clock is modified by placing it in a gravitational field. This is something that was not at all anticipated by the geometers of the nineteenth century, nor by Schwarzschild, as far as I can tell.

[1] Cited in H. P. Robertson, "Geometry as a Branch of Physics," in Paul Arthur Schilpp, ed., *Albert Einstein: Philosopher-Scientist*, p. 323.

To put the matter baldly, a clock in a gravitational field is "slow" when compared to an identical clock free of the effects of gravity, and the stronger the gravity is in a given region the slower the clock will be. We may get a feeling for how this works by considering a clock attached to the rim of a rotating wheel and comparing its reading to the reading of an identical clock at rest at the center of the wheel. Suppose, for the sake of illustration, we take as our clock the lifetime of some decaying elementary particle. As we have seen, this lifetime is increased as compared to the lifetime of the same particle when at rest. The more rapidly we spin the wheel the longer will our particle appear to live when we view it from the center of the wheel. To an observer attached to the wheel the particle will appear at rest. But he will be subject to a force pushing him away from the wheel. By the principle of equivalence he can regard this as a gravitational force attracting him away from the wheel. Hence he will say that the lifetime is longer because he is in a gravitational field while the observer at the center of the wheel is in a region of space free of gravity. The general theory of relativity yields the same result for clocks of arbitrary construction subjected to arbitrary gravitational forces.

In fact in Einstein's 1916 paper he makes mention of astronomical data that he thought might confirm such a prediction. He had in mind what is now known as the "gravitational red shift." Atomic electrons when they jump from one orbit to another give off light with characteristic frequencies—i.e., colors. These "spectra" are used to identify chemical substances, and, in particular, spectral lines have been used, since the mid-nineteenth century, to identify the elements in stars and, of course, the sun. These atomic vibrations are also a kind of clock, and according to the general theory of relativity they should slow down when the atoms are in the strong gravitational fields at the surface of stars. Hence the light should be shifted to the red, since red light has a smaller frequency than, say, blue light. Einstein thought this effect might already have been seen at the time he wrote his paper. We now know that separating this effect from the general background of complicated astronomical effects that change the spectral characteristics of starlight is much more difficult than first imagined, and it is only in the last few years that astronomers finally believe that they really *have* seen the Einstein red shift in stars. Meanwhile, some extremely beautiful experiments done on *earth* by independent groups at Harvard University (led by Professor R. V. Pound) and at Harwell, in England, have measured the Einstein effect in a manner that seems to leave no doubt about it.

At Harvard, there is a 74-foot tower in the Jefferson Laboratory—which is to say that the bottom of the tower is 74 feet closer to the center of the earth than the top and hence the gravitational field at the bottom is slightly stronger than at the top. This difference produces, according to the Einstein theory, an incredibly small light shift (one part in 10^{15}). Until a few years ago the detection of such a small frequency shift would have been beyond experimental technology. But new methods in quantum optics have now made it possible, and these were employed by Pound and his co-workers and by the Harwell group. In the interests of accuracy we should remark that what the experiment of Pound and his collaborators detects is not a red shift but rather a blue shift of the light. This is actually not so difficult to understand: if we think of the light as being made up of quanta, then as these "fall" down the tower toward the bottom they *gain* energy, since, like a ball that is dropped toward the earth, they move into a region of stronger gravitation. As will be explained in the next chapter a gain in energy means an increase in the frequency of the light and hence a shift to the blue, since blue light has a higher frequency than, say, red light. On the other hand light leaving the sun and moving to the earth *loses* energy, since the earth has a weaker gravitational field than the sun and hence such light is red-shifted. An inhabitant on the sun who performed experiments on light coming from the earth would find this light blue-shifted, as in the Pound experiment, since it is moving from a weak to a strong field of gravity. In these experiments Pound and his collaborators have been able to confirm Einstein's prediction to an accuracy of about one per cent. The essential trick that makes these experiments possible is this: nuclei can emit photons of a precisely defined energy when the nuclei make transitions from one state to another. These photons may be reabsorbed by similar nuclei to reverse the transition. This process can be affected if, as in the Harvard experiment, the light undergoes a change in wave length while it is traveling from the emitter to the absorber. The absorber is no longer in "resonance" with the emitter, and the absorption is reduced, which demonstrates that the light frequency has been shifted. The real difficulty that these experiments must overcome is that the nuclei in matter tend to jiggle around, and this jiggling leads to a lack of precision in defining the light frequency. In 1958 the German physicist R. L. Mössbauer discovered a method of implanting these nuclei in crystals, which eliminated the jiggling. This "Mössbauer effect" is at the heart of modern methods to measure the red shift.

In his 1916 paper Einstein made two predictions—the bending of light by the gravitational field of the sun and the red shift, both of which have been confirmed— and a deduction. The deduction solved a problem in planetary astronomy that had puzzled astronomers from at least the middle of the nineteenth century. It was well known to astronomers, from the time of Newton, that the sun's gravitational field is not the only influence on a planet's motion. The planets influence one another's motions to some degree by *their* mutual gravitation. At first it looked as if these "perturbations"—although small—might add up in such a way that the whole system was unstable, so that eventually the solar system, or parts of it, could become unstuck and the planets simply wander off into space, or perhaps crash into the sun. However, at the end of the eighteenth century Laplace proved that a special feature of Newton's law was that this could not happen, even though the perturbed orbits were not simple closed ellipses. In 1845, the French astronomer Leverrier found by an actual analysis of the observed orbit of Mercury, the planet closest to the sun, that its orbit did *not close*. Each time the planet went once around the sun it returned to a different point in space so that the orbit, if viewed long enough, would not be a simple ellipse, but many neighboring ellipses which if superposed would look something like the petals of a flower. The way in which this effect is usually demonstrated is by picking the point closest to the sun on the orbit—the so-called perihelion—and seeing how it changes in time as the years pass. It turns out that the perihelion of mercury changes by the rather minute, but measurable, amount of about .43. seconds of arc per *century* over and above what could be accounted for by Newtonian gravitation, taking into account all of the perturbing effects of the known planets. For a while astronomers thought there might be a hidden planet (which Leverrier named Vulcan) near the sun that caused this effect, but by the beginning of the twentieth century no such planet had been found and astronomers were beginning to suspect Newton's law of gravitation.

It is a remarkable feature of Einstein's theory of gravitation that it predicts exactly this perihelion advance for Mercury. It also predicts a similar effect on the orbits of the other planets, but since they are farther away from the sun the effect is much smaller. A footnote to this, which, no doubt, would have pleased Einstein, is that in 1968 the American physicist I. I. Shapiro and his collaborators, using radar techniques, were able to plot the orbit of the asteroid Icarus and found that it had a perihelion advance in general conformity to the theory. (Shapiro and his group have also been bouncing radar signals off such planets as Mercury, when these planets are lined up so that the signals graze the sun, in order to make a new test of Einstein's prediction that electromagnetic radiation—light, radar, and so on—is bent by the gravitation of the sun. Here too the theory is in general agreement with experiment.)

An even more profound aspect of the geometry of gravitation suggested by Einstein's theory is the new insights it gave to the ancient question of cosmology.

The word "cosmology" is derived from the Greek words *kosmos*—meaning "the world," i.e., the universe —and *logos*—meaning "discourse": "discourse" that is concerned with the structure and evolution of the universe as a whole. Because of the daring and apparently hopelessly difficult questions that cosmology tries to answer and the paucity of experimental data, it has been, until recently, regarded as something of a black sheep in the scientific family, more like science fiction than science. At the moment, largely because of new astronomical discoveries, it is one of the most exciting and active branches in *all* of science. This change can be attributed, at least in substantial part, to developments that stem from a paper that Einstein published in 1917— "Cosmological Considerations on the General Theory of Relativity"[2]—which, curiously enough, is now "antiquated if not even wrong." Leopold Infeld, who wrote this in 1947, added, "I believe Einstein would be the first to admit this. . . . Yet the appearance of this paper is of great importance in the history of theoretical physics. Indeed, it is one more instance showing how a wrong solution of a fundamental problem may be in-

[2] The original paper, "Kosmologische Betrachtungen zur allgemeinen Relativitätstheorie," is translated and reprinted in *The Principle of Relativity*, pp. 177–88.

comparably more important than a correct solution of a trivial, uninteresting problem."[3]

Einstein begins his paper with a "critique" of the cosmology that would be suggested by a straightforward application of Newton's theory of gravitation. We have put "critique" in quotation marks because, while Einstein's remarks are logically consistent, the Newtonian picture correctly understood, as opposed to what Newton did, is actually closer to what we, and later Einstein himself, now regard as the real cosmological situation. Newton made an attempt to construct a universe of stars, extending to infinity, in which the stars would be uniformly distributed, on the average, throughout the universe. These stars were supposed to interact with one another through the effects of their mutual gravitation. Newton wanted to explain what he took to be a static and unchanging universe of infinite extent. What Einstein pointed out, and this had been noticed by others as well, is that given Newton's law of gravitation this situation cannot exist. We now know—and we will discuss how we know this later—that in fact this situation *does not* exist and that the universe is, at present, *expanding*. If Newton had been so disposed he might have predicted a universe that evolves in space rather than trying, as he did, to account incorrectly for a static universe. However, in 1917, all of the astronomical evidence still suggested a static universe, so Einstein was led to try to fit such a picture into *his* theory of gravitation. He soon realized that the same objections that he leveled against Newton were equally valid when applied to his own theory—so, in fact, he modified his theory. This is the subject of his 1917 paper. As he wrote:

> I shall conduct the reader over the road that I have myself traveled, rather a rough and winding road, because otherwise I cannot hope that he will take

[3] Leopold Infeld, *Albert Einstein: His Work and Influence on Our World*, p. 72.

much interest in the result at the end of the journey. The conclusion I shall arrive at is that the field equations of gravitation which I have championed hitherto still need a slight modification, so that on the basis of the general theory of relativity those fundamental difficulties may be avoided which have been set forth . . . as confronting the Newtonian theory.[4]

To his old gravitational equation he added a new term—called the "cosmological term"—characterized by a new small constant (in addition to the gravitational constant). This "cosmological constant" does not modify the old predictions of the theory that have to do with "local" phenomena involving the solar system. Einstein was able, with the help of the mathematician J. Grommer, to find a solution to these equations which appeared to be static. (Sometime later Eddington showed that, in fact, the solution was not really static in the sense that if the universe received any slight jiggle anywhere it would start to expand or contract.) This solution pictures the universe as a sort of sphere in space, filled, on the average, with a small uniform density of matter. A light ray started at any point in the universe will return to its starting point, according to this picture, in about 10 billion years. At about the same time the Dutch astrophysicist Willem de Sitter found a second solution to the equations which corresponded to the somewhat artificial situation in which the average matter density is zero, but which did have the interesting property that this de Sitter universe was expanding. This was the first theoretical suggestion of an expanding universe.

The next important step came in 1922 when the Russian scientist (he worked in several fields) A. Friedmann dropped the cosmological term, reverting to Einstein's original equations, and found in fact that they had a solution for an expanding universe with a

[4] *The Principle of Relativity*, pp. 179–80.

non-zero matter density. At first Einstein did not believe this result and even attempted a refutation which found its way into print.[5] He soon realized that Friedmann's work was correct[6] and, along with most other cosmologists, abandoned the cosmological term as an extra and unnecessary complication. Most of modern cosmological thinking makes use of the Friedmann solutions (there are more than one). By 1929, following the work of the American astronomer E. P. Hubble, primarily, it had become clear that the universe *was* expanding. This is such a significant fact that it is worth commenting on.

Until early in the twentieth century there was no firm evidence that any of the astronomical objects which astronomers had observed (such as nebulae) lie outside our own Milky Way galaxy. (By "nebulae" we mean collections of stars bunched together which look like luminous clouds until the individual stars are resolved by a large telescope. There are also gas clouds in interstellar space which astronomers, properly so, call nebulas. Our "nebulae" are in reality "galaxies.") But with the work at Harvard in 1912 of Henrietta Leavitt and, later, Harlow Shapley, it was shown that these objects were in fact outside the Milky Way. (Determining astronomical distances for really distant objects is a very tricky business which would take us too far afield to go into here. Suffice it to say that in 1952 the astronomer W. Baade made a new analysis of Shapley's methods and discovered, to everyone's amazement, that all of these large distances were at least *twice as large* as what Shapley had thought.) The next important step was made by V. M. Slipher, beginning in 1912 at the Lowell observatory, who observed that these distant galaxies showed a "red shift" of their light. This means that the spectral lines from the starlight of

[5] *Zeitschrift für Physik*, 11 (1922), 326.
[6] *Ibid.*, 16 (1923), 228.

these galaxies is displaced toward the red as compared to the nearer stars. By 1929 Hubble had shown that this red shift obeyed a very simple law. The farther away the galaxy, the more it is red-shifted, and in fact the red shift is simply proportional to the distance. We know that light emitted by an object that is moving away shows a Doppler red shift just as sound emitted by a moving train has its pitch altered. Hence it was, and is, natural to assume that distant astronomical objects are receding from us in such a way that the farther away an object is the faster it is receding. Some very distant astronomical objects have been observed in which there are red shifts of more than half the normal wave length.

In 1927 the Belgian abbé Georges Lemaître introduced the idea that the expansion began with a gigantic cosmic explosion—the "big bang" in the terminology of the late George Gamow—which took place some 10 billion years ago, or so, when the matter of the universe was compressed into a state of extremely high density. Gamow and his collaborators and successors have been working for the last thirty years, with considerable success, in trying to explain the observed distribution of matter and radiation in the universe via the nuclear physics of this explosion. (It is, perhaps, interesting to point out that nuclear physics was one of the few branches of modern physics to which Einstein did not contribute, no doubt because when it began, in the modern sense, in the 1930s Einstein was almost entirely preoccupied with his unified field theories.) It now seems as if the *quasars*—quasi-stellar radio sources—which, if present notions hold up, are distributed non-uniformly outside our galaxy, may be remnants of this primeval explosion along with the recently discovered cosmic "black body" radiation which seems to fill the universe and may reflect an aftereffect of the original explosion. The over-all geometry of the universe is not yet completely known, but there is reason to believe that it may be a Friedmann geometry with positive curvature like a sphere and with the property that light emitted at any given point will eventually come back. This is what is meant by saying that the universe is "closed." Needless to say, these sketchy remarks do not do even partial justice to modern cosmology, but they may be enough to indicate the richness of the ideas contained in the general theory of relativity.

3. The Quantum Theory

Preamble: Einstein and Newton

• • •
XIII

Historians of science often refer to the year 1666 as the *annus mirabilis* of classical science. It was during this year that Isaac Newton, returned to the protection of his mother's house in Lincolnshire by the Great Plague of 1665 that had closed Cambridge University, formulated— at least for himself (much remained unpublished for decades)—most of the basic concepts that transformed physics into a serious quantitative science. He was then twenty-four and was, as he later reminisced, "in those days . . . in the prime of my age for invention and minded mathematicks and philosophy [natural science] more than at any time since."[1] During a period of eighteen months Newton formulated his basic laws of mechanics and the calculus, differential and integral, for working out their consequences, as well as the law of universal gravitation and

[1] Cited in Frank E. Manuel, *A Portrait of Isaac Newton*, (Cambridge, Mass., 1968), p. 80. Not all scholars agree with Professor Manuel's brilliant portrait, but I find it convincing.

his optical discoveries, of which the most celebrated is his observation that "white light" from the sun is dispersed by a prism into a rainbow of colored light, each color being bent by the prism at a different and characteristic angle.

As a physicist Newton had every virtue. He was a careful and imaginative experimenter, a mathematical genius of the first magnitude, and he had the capacity to focus his powers on a single problem day after day, for years if necessary, until he was able to find and formulate the solution in full generality and with lucid elegance. In 1931 Einstein wrote a foreword to a new edition of Newton's *Opticks* in which he commented:

> Fortunate Newton, happy childhood of science! He who has time and tranquility can by reading this book live again the wonderful events which the great Newton experienced in his young days. Nature to him was an open book, whose letters he could read without effort. The conceptions which he used to reduce the material of experience to order seemed to flow spontaneously from experience itself, from the beautiful experiments which he ranged in order like playthings and describes with an affectionate wealth of detail. In one person he combined the experimenter, the theorist, the mechanic and, not least, the artist in exposition.[2]

It would be pleasing to be able to report that Newton as a human being matched, in some sense, the overwhelming grandeur of his scientific creation. But this is not the case. In fact Newton—a fanatical puritanical man who seems to have died a virgin and, probably, had suppressed homosexual tendencies—could be vicious and petty, given to violent rages (which he was able fully to satisfy during his tenure as Warden of the Mint when he held the power of life and death over the wretched "coiners and clippers"—counterfeiting of coinage was then a capital offense in England—who came

[2] Cited in *ibid.*, p. 68.

before his judgment). He spent at least as much time, perhaps more, on alchemy and Biblical chronology as on physics. And he appears to have been almost totally unscrupulous in the pursuit of his personal ambitions, which often had more to do with his position in society than anything else. Part of the Newtonian miracle is that despite these psychological handicaps—maybe even because of them—he was able to channel his intellectual and psychic energies with such intensity and even ferocity on scientific discovery.

> We stand in awe before genius, and wonder the more because its works are achieved in the face of odds that have crushed lesser men. When the magnitude of the forces that might have led to disintegration is recognized, the acts of sheer will and overcoming astound. There have doubtless been others with psychic configurations similar to Newton's who have never been heard of since.[3]

We are still too close to the events to be as confident in our judgments about Einstein's achievements as we can be about Newton's, but almost every contemporary physicist would agree that the year 1905 can be regarded as the *annus mirabilis* of modern physics. Each of Einstein's four papers published in the *Annalen der Physik* in 1905 has resulted, in quite different ways, in changing our views of the physical universe. Two of the papers, those on the special theory of relativity, form a unit in terms of style and subject matter; one had to do with the Brownian motion—the incessant motion of suspended microscopic particles in liquids; and the last (actually the first to be published) was the foundation of quantum physics. In 1949 Max Born expressed with eloquence and authority what every physicist has come to feel about this work:

> One of the most remarkable volumes in the whole of scientific literature seems to me Volume 17 (4th

[3] *Ibid.*, pp. 2–3.

Series) of *Annalen der Physik*, 1905. It contains three papers by Einstein . . . each today acknowledged to be a masterpiece, the source of a new branch of physics. These three subjects, in order of pages, are: theory of photons, Brownian motion, and relativity. . . .

Relativity is the last one, and this shows that Einstein's mind at that time was not completely absorbed by his ideas on space and time, simultaneity and electro-dynamics. In my opinion he would be one of the greatest theoretical physicists of all times even if he had not written a single line on relativity—an assumption for which I have to apologize, as it is rather absurd. For Einstein's conception of the physical world cannot be divided into watertight compartments, and it is impossible to imagine that he should have by-passed one of the fundamental problems of the time.[4]

No one who is really competent has attempted as yet to make something like a complete psychological or psychoanalytical portrait of Einstein, and it is only in the last few years that scholars have begun to sift through the enormous collection of documents—letters, unpublished manuscripts, and the like—that are becoming available.[5] Whatever we may learn from these documents, it is already clear that they contain nothing of the startling character of the "Portsmouth Papers"— the contents of Newton's celebrated chest of secret docu-

[4] Max Born, "Einstein's Statistical Theories," in Paul Arthur Schilpp, *Albert Einstein: Philosopher-Scientist*, p. 163.
[5] In addition to hundreds of unpublished letters, there are "eleven notebooks starting from Einstein's student days; a few travel diaries; folders upon folders of published manuscripts, many in early draft; and several dozen unpublished manuscripts. All this survived, more or less by good luck; on returning from a trip to the United States in the winter of 1932–33, Einstein found on reaching Europe that Hitler's supporters had taken over Germany. Einstein never again set foot on German soil, and most of the correspondence was brought out by diplomatic pouch through the French Embassy." Gerald Holton, "Influences on Einstein's Early Work in Relativity Theory," pp. 62–63.

ments which he assembled in 1696 when he left Cambridge for London. The chest passed ultimately into the possession of John Maynard Keynes—nearly one million words of handwritten manuscript on alchemy, Church history (Newton was, in secret, a Unitarian in a time when this was considered a grave heresy), and apocalyptic Biblical writing, which to Newton was to be regarded as scientific evidence on the same basis as the revelations of his prism—and Keynes appears to have been the first modern scholar to appreciate something of what Newton was really like:

> In the eighteenth century and since, Newton came to be thought of as the first and greatest of the modern age of scientists, a rationalist, one who taught us to think on the lines of cold and untinctured reason. I do not see him in this light. I do not think that any one who has pored over the contents of that box which he packed up when he finally left Cambridge in 1696 and which, though partly dispersed, have come down to us, can see him like that. Newton was not the first of the age of reason. He was the last of the magicians, the last of the Babylonians and Sumerians, the last great mind which looked out on the visible and intellectual world with the same eyes as those who began to build our intellectual inheritance rather less than 10,000 years ago. Isaac Newton, a posthumous child born with no father on Christmas Day, 1642, was the last wonderchild to whom the Magi could do sincere and appropriate homage.[6]

While Einstein is the only modern scientist one can begin to compare, from point of view of achievements, with Newton, it is difficult to find very much that they had in common as men. Everyone who had real contacts with Einstein came away with an overwhelming sense of the nobility of the man. The phrase that recurs

[6] J. M. Keynes, "Newton the Man," reprinted in Henry A. Boorse and Lloyd Motz, eds., *The World of the Atom*, pp. 93–101.

again and again is his "humanity"—or, as trite as it may sound, the simple, lovable quality of his character. Nowhere in all of his professional life is there the remotest sense of the often bitter competitiveness, the struggle over claims to scientific invention, that cloud and sometimes destroy the lives of scientists. Of course one can say that with his achievements he had no need for such things, and it could also be said that Einstein, unlike Newton, lived in an era in which physics had become a recognized professional activity with an implicit "gentlemanly" code of ethics. (Most of Newton's scientific contemporaries, judging from their correspondence, appear to have been nearly as unpleasant as Newton.) But this is beside the point. Newton had no "need" for such things either—in the sense that his scientific reputation was never in jeopardy. It was universally recognized by the end of the seventeenth century, long before his death, that he was among the greatest scientists who had ever lived. Yet he was at odds—the most savage kind of personal warfare—with every one of his contemporaries who were anything like his intellectual equals and who insisted on being treated as such. Newton had either sychophantic followers who treated him like a prima donna, or bitter rivals. There were no collaborators. Einstein had serious scientific differences with most of his scientific contemporaries—especially later in his life, when he rejected quantum theory—but in his dealings with them there is always a sense of mutual respect and, on Einstein's side, real pleasure on learning about new discoveries or new ideas. It is completely typical that it was Einstein who first recognized the significance of the ideas on the wave nature of matter propounded, without a shred of direct experimental evidence, in 1924 in the Ph.D. thesis of the then unknown French physicist Louis de Broglie. Later Einstein came to reject the usual interpretation of the "de Broglie waves," but, as de Broglie wrote, "The scientific world of the time hung on every one of Einstein's words, for he was then at the peak of his fame. By stressing the importance of wave mechanics, the illustrious scientist had done a great deal to hasten its development. Without his paper my thesis might not have been appreciated until very much later."[7]

Among the things that Einstein and Newton seem to have had in common apart from their genius—whatever that means—was an all but incredible capacity for single-minded concentration on a scientific problem. This is something that is quite different than a capacity for "hard work," which they both had also. Hard work, in the conventional sense, for a theoretical physicist means doing long and sometimes very tedious computations. This requires, in my opinion, less in the way of actual concentration than does the grasping of a problem in one's mind in such a way that it can be reduced to a computation. Theoretical physicists are often fond of saying that formulating the correct question is much more difficult than trying to answer it. The really deep discoveries in theoretical physics do not begin as well-posed problems. A well-posed problem implies the existence of a theory that provides the language in which to state the questions. Physics textbooks are full of well-posed problems that sometimes require great ingenuity to work out, but that are not really "deep." They require a perhaps unlikely juxtaposition of the formulas and ideas in the book. But nature is not a textbook, and while one can trace the antecedents of a discovery like Newton's law of universal gravitation, or like the special theory of relativity, there is some kind of mental leap involved in making such a discovery that carries it beyond its antecedents, often in wholly unanticipated new ways. In his essay on Newton Lord

[7] Cited by Martin J. Klein, "Einstein and the Wave-Particle Duality," in *The Natural Philosopher*, 3(1964), p. 38. Professor Klein, of Yale University, had made an extensive study of Einstein's contributions to, and attitudes toward, the quantum theory.

Keynes describes Newton's powers of concentration in a manner that applied essentially verbatim to Einstein:

> I believe that the clue to his mind is to be found in his unusual powers of continuous concentrated introspection. A case can be made out, as it also can with Descartes, for regarding him as an accomplished experimentalist. Nothing can be more charming than the tales of his mechanical contrivances when he was a boy. There are his telescopes and his optical experiments. These were essential accomplishments, part of his unequalled all-round technique, but not, I am sure, his *peculiar* gift, especially amongst his contemporaries. His peculiar gift was the power of holding continuously in his mind a purely mental problem until he had seen straight through it. I fancy his pre-eminence is due to his muscles of intuition being the strongest and most enduring with which a man has ever been gifted. Anyone who has ever attempted pure scientific or philosophical thought knows how one can hold a problem momentarily in one's mind and apply all one's powers of concentration to piercing through it, and how it will dissolve and escape and you find that what you are surveying is a blank. I believe that Newton could hold a problem in his mind for hours and days and weeks until it surrendered to him its secret. Then being a supreme mathematician he could dress it up, how you will, for purposes of exposition, but it was his intuition which was pre-eminently extraordinary—"so happy in his conjectures," said de Morgan, "as to seem to know more than he could possibly have any means of proving." The proofs, for what they are worth, were as I have said, dressed up afterwards—they were not the instrument of discovery.[8]

Like Newton Einstein could and did concentrate on individual problems for years at a time.[9] The special

[8] Keynes, *op. cit.*, p. 95.
[9] Einstein once remarked to an assistant, "God is inexorable in the way He has allotted His gifts. He gave me the stubbornness of a mule and nothing else, really. He also gave me a keen scent." Quoted in Klein, *op. cit.*, p. 46.

theory of relativity required, from all accounts, nearly a decade of preparatory thought, although, as he later remembered it, the final formulation and the writing of the manuscript took only five or six weeks. The general theory of relativity and gravitation took about seven years to complete allowing for all of the false starts, and Einstein worked constantly on the unified field theory— an attempt to unite gravitation and electromagnetism —despite the critical opposition of most of his contemporaries, who were as convinced that he was on the wrong track as he was convinced that he was not, for more wrong track as he was convinced that he was not, for more than three decades. Moreover, the early work of Einstein—until the papers on the general theory of relativity—have just this incredible "intuitive" sense that Keynes sees in Newton. There is a major difference in presentation in that Newton, above all in the *Principia*, adopts the rigorous impersonal formal style of a geometry text. In reading the early papers of Einstein one has—perhaps erroneously—the sense of being close to the thinking processes of the man. They are full of such phrases as, "In a memoir published four years ago I tried to answer the question whether the propagation of light is influenced by gravitation. I return to this theme, because my previous presentation of the subject does not satisfy me. . . ."[10] We have the continual sense that these papers have been written by a human being, and that we are witness to his "personal struggle" with the puzzles and mysteries of the natural universe.

There is also the strikingly nonmathematical character of these early papers. There are relatively few equations—often not even numbered sequentially. (Not a single equation in the 1905 paper on special relativity is numbered, in contrast to the 1916 paper on the general theory of relativity, where there are seventy-five numbered equations.) These papers are full of *ideas* ex-

[10] From Einstein's 1911 paper on gravitation. See *The Principle of Relativity*, p. 99.

pressed verbally, and, often, in terms of simple thought experiments which enable one to visualize what is going on prior to and concomitant with the mathematical expression. Einstein was, apparently, not a particularly good mathematical calculator and he did not find his results as a consequence of long calculations. (He often claimed that his memory was bad, and prodigious calculational ability is usually combined with an exceptional memory.) He found his results by a phenomenal intuitive instinct as to what the results *should* be. Given our general ignorance about the workings of the human mind it is difficult to imagine that anyone would attempt to "explain" the creativity of someone like Einstein. But Einstein himself tried at various times at least to describe what appeared to him to be distinctive about his thought processes, and what stands out is the role of pictures—visual imagery—as opposed to words.

What, precisely, is "thinking?" When at the reception of sense-impressions, memory pictures emerge, this is not yet "thinking." And when such pictures form series, each member of which calls forth another, this too is not yet "thinking." When, however, a certain picture turns up in many such series, then—precisely through such return—it becomes an ordering element for such series, in that it connects series which in themselves are unconnected. Such an element becomes an instrument, a concept. I think that the transition from free association or "dreaming" to thinking is characterized by the more or less dominating role which the "concept" plays in it. It is by no means necessary that a concept must be connected with a sensorily cognizable and reproducible sign (word); but when this is the case thinking becomes by means of that fact communicable. . . .

With what right—the reader will ask—does this man operate so carelessly and primitively with ideas in such a problematic realm without making even the least effort to prove anything? My defense: all our thinking is of this nature of a free play with concepts;

the justification for this play lies in the measure of survey over the experience of the senses which we are able to achieve with its aid. The concept of "truth" cannot yet be applied to such a structure; to my thinking this concept can come in question only when a far-reaching agreement (*convention*) concerning the elements and rules of the game is already at hand. . . .

For me it is not dubious that our thinking goes on for the most part without the use of signs (words) and beyond that to a considerable degree unconsciously. For how, otherwise, should it happen that sometimes we "wonder" quite spontaneously about some experience? This "wondering" seems to occur when an experience comes into conflict with a world of concepts which is already sufficiently fixed in us. Whenever such a conflict is experienced hard and intensively it reacts back upon our thought world in a decisive way. The development of this thought world is in a certain sense a continuous flight from "wonder."[11]

Anyone who has ever worked on a scientific problem has experienced such flights from "wonder." If the scientific discipline is at the stage of development where a new and significant breakthrough has been made—say atomic physics in the 1930s after the discovery of modern quantum theory—then what is involved is the reduction of phenomena to the newly created theoretical mold: a form of problem-solving not too far removed from the textbook exercise. However, the creation of the new theory itself appears to require a different sort of reduction altogether. Here the creator must fit the phenomena into a pattern that he alone can discern. This pattern is shaped by some sort of "intuitive" sense —a "vision"—of what the world should be like. As different as they apparently were as men, both Newton and Einstein shared a feeling in the fitness of their own intuitions. As Einstein once put it, "To him who is a dis-

[1] Schilpp, *op. cit.*, pp. 7–8.

coverer in this field the products of his imagination appear so necessary and natural that he regards them, and would like to have them regarded by others, not as creations of thought but as given realities."[12] When a certain point in the creative process had been reached, they "knew" that they must be right and, at least in Einstein's case, could wait patiently and confidently for experiment to bear him out. I am not aware of any example in which Newton's intuitions in physics were seriously frustrated by experimental developments which occurred during his lifetime. For much of Einstein's creative life the same thing was true of him as well. However, we find in Einstein's reaction to the quantum theory a sharp break between his sense of the fitness of things and what the experimental discoveries appeared to be revealing. As we shall see, his reaction to this was to withdraw deeper and deeper into the solitude of his own intuitions.

[12] *Essays in Science*, p. 12.

Brownian Motion

•

XiV

No physicist had more to do with the creation of quantum physics than Einstein. His work in this field would have been in and of itself a full scientific career for any other physicist. Yet just when, after nearly thirty years, physicists succeeded in changing what had appeared to be an intellectual chaos into a logical scientific structure—a structure that many people feel is the supreme triumph of twentieth-century scientific thinking—Einstein turned his back on it. Einstein wrote extensively, both in correspondence and articles, explaining his reasons for his rejection of the theory, despite the fact that, in Dirac's phrase, it "accounts for all of chemistry [we might now be tempted to add biology] and most of physics." In reading these explanations one is inevitably struck by two things. First, there is the man's incredible stubbornness. He would not be budged. He listened to arguments against his position for nearly thirty years, listened and absorbed, and refuted, and would not budge. The second thing one is struck by is his ultimate reason for rejecting the theory: the phrase that occurs again and again is "incompleteness." He could not accept in a fundamental physical theory a description that gave up "visualizable", causal events for probabilities. In some sense he remained rooted all his life in the visual and geometric sensibilities that belonged among his most vivid childhood memories. We shall return to these themes in what follows. But we must first trace the path that leads up to them, and this, naturally, takes us back to 1905 and Einstein's two other masterworks published that year in the *Annalen der Physik*.

Of Einstein's great 1905 papers the most accessible one, at least from the point of view of the significance of its content, is the one on the so-called Brownian motion.[1] It was the experiments suggested by this paper that convinced the skeptics, and there were many even in 1905, of the existence of atoms. The statistical techniques that Einstein developed to cope with this atomic problem he also made use of, at about the same time, to study the quantum aspects of radiation. Hence, the quantum-theory and Brownian-motion papers of 1905 are, in fact, applications of a unified idea, namely, the mathematical description of systems involving enormous numbers of basic units—atoms or light quanta—by statistics. Since we are now convinced of the existence of atoms, the Brownian motion is the simplest place to begin the discussion of this aspect of Einstein's work.

[1] Einstein's papers on Brownian motion have been translated and collected in R. Fürth, ed., *Albert Einstein: Investigations on the Theory of the Brownian Movement*. The full title of the 1905 paper (the first in the volume) is "On the Movement of Small Particles Suspended in a Stationary Liquid Demanded by the Molecular Kinetic Theory of Heat."

Robert Brown was a Scottish botanist who, in the summer of 1827, made the experiments that now bear his name. The basic observation was exceedingly simple. Brown studied, through an ordinary microscope, the behavior of pollen grains—particles from various plants which in the original experiments measured something like 1/5,000 of an inch—when immersed in water. What he discovered was that these particles perform a constant, agitated, and apparently erratic motion that has nothing to do with any currents moving in the water. As Brown put it, "These motions were such as to satisfy me, after frequently repeated observation, they they arose neither from currents in the fluid, nor from its gradual evaporation, but belonged to the particle itself."[2]

In this Brown made the natural, albeit misguided, assumption that his microscopic particles represented some sort of new state of matter, to which he gave the name "active molecule." (At first he thought that they might be alive, but he repeated the experiment with dried pollen particles of plants that had been preserved in a herbarium for "upwards of twenty years.") Pretty soon he extended his investigations to nonvegetable substances—microscopic particles of gum resin, coal tar, manganese, nickel, plumbago, bismuth, antimony, and arsenic, among others. And, as he wrote, the active "molecules were found in abundance."[3] In other words microscopic particles of *anything* suspended in

[2] A brief description of Brown and his work can be found in Henry A. Boorse and Lloyd Motz, eds., *The World of the Atom*, I, 206–12. This quotation, taken from Brown's original article in the *Philosophical Magazine* (1828), is given on page 209.

[3] *Ibid.*, p. 210. Brown goes on to add, "I remark here also, partly as a caution to those who may hereafter engage in the same inquiry, that the dust or soot deposited on all bodies in such quantity, especially in London, is entirely composed of these molecules."

water—or any liquid—carry out constant, random, agitated motions which are now called "Brownian" motions. While this phenomenon was frequently studied in the nineteenth century (in 1865 one group of investigators even showed that the motion continued unabated for an *entire year* while they carried out their experiment on a liquid sealed in glass) it was not until a half century after Brown's discovery that several scientists made the qualitatively correct suggestion as to what Brownian motion was. They noted that if the liquid itself were taken to be composed of molecules —these are not the "active molecules" of Brown but the molecules in the sense of the modern chemist: the smallest units of the matter composing the liquid which exhibit its chemical properties—and if these molecules were in a state of constant random agitation, then the suspended particles would reflect this agitation, since they would be continually bombarded, from all sides and in all directions, by the molecules of the liquid.

To us, as accustomed as we are to the atomic theory of matter, this explanation may seem almost self-evident. But, while the atomic hypothesis had gained considerable support among physicists and chemists, at the end of the nineteenth century it was still that—an hypothesis; its universal validity was neither clear nor completely accepted. Moreover, the connection with the Brownian motion was still vague, and the experiments still uncertain. At the beginning of his 1905 paper Einstein wrote:

> In this paper it will be shown that according to the molecular-kinetic theory of heat, bodies of microscopically-visible size suspended in a liquid will perform movements of such magnitude that they can be easily observed in a microscope, on account of the molecular motions of heat. It is possible that the movements to be discussed here are identical with the so-called "Brownian molecular motion"; however, the information available to me regarding the latter is so lack-

ing in precision, that I can form no judgment in the matter.[4]

The long history of atoms begins, of course, with the familiar statement of Democritus, in the fourth century B.C.: "The only existing things are the atoms and the void; all else is mere opinion." If taken too literally, such a statement is certainly nonsense. But the significance of it—and this is really the great contribution of the Greek atomists—was the idea that behind the bewildering complex appearance of the forms of matter lay an underlying structure of "atoms"—indivisible particles—obeying simple laws that enable us to explain and correlate the experience of our senses. The conceptual nature of these atoms has changed in the past two thousand years—we have become more sophisticated and the phenomena to be explained are more subtle—but the atomic ideal is still with us. In its latest manifestation there are the so-called elementary particles out of which all matter is to be composed. But by now high-energy physicists have found so many "elementary particles" that one has begun to look for an underlying substratum of *super*elementary particles out of which even the old-fashioned elementary particles such as the neutron, proton, and the several mesons are composed. Since the evidence for these *super*elementary particles is, as yet, entirely indirect—they allow a simplification in theoretical description ("direct" evidence would be the actual production of such particles in high-energy accelerators)—modern physicists find themselves in the ironic position of the nineteenth-century physicists and chemists who were unsure of the "existence" of atoms.

Newton appears to have been an atomist. As he wrote in the *Opticks*:

It seems probable to me, that God in the Beginning form'd Matter in solid, massy, hard impenetrable, moveable Particles of Such Sizes and Figures, and

[4] *Theory of the Brownian Movement*, p. 1.

with such other Properties and in such Proportion to Space, as most conduced to the End for which he form'd them; and that these primitive Particles being Solids, are incomparably harder than any porous Bodies compounded of them; even so very hard, as never to wear out or break in pieces; no ordinary Power being able to divide what God himself made one in the first Creation.[5]

Indeed, such support as existed for the atomic hypothesis had, until the beginning of the eighteenth century, this sort of theological and metaphysical origin. Essentially nothing of a quantitative character had resulted for science from it. However, in 1738 Daniel Bernouilli, a member of a remarkable family of Swiss mathematicians, published a book called *Hydrodynamica* which contained the first example of what we would now call the statistical mechanical approach to the physics of gases, in which the atomic hypothesis was used quantitatively. He gave a demonstration of what is known as Boyle's law for gases. This was the experimental observation by Newton's contemporary Robert Boyle that if a gas is maintained at a given temperature in a container, the pressure on the container increases as its volume is shrunk in such a way that the pressure increases in proportion to the shrinking volume. (If one squeezes a balloon filled with air it will eventually break, since the pressure increases as the volume of the balloon is shrunk.) This Bernouilli explained by considering the gas as being made up of small "corpuscles, which are driven hither and thither with a very rapid motion. . . ." The pressure, in this picture, is caused by collisions of the gas molecules with the walls of the container, and by analyzing how these increase in frequency as the volume of the container shrinks, Bernouilli derived Boyle's law.[6]

[5] From the *Opticks*, cited in Boorse and Motz, *op. cit.*, p. 102.
[6] The details of Bernouilli's derivation are given in Boorse and Motz, *op. cit.*, pp. 112–16.

No additional progress in the kinetic theory of gases was made until the middle of the next century, and in the meanwhile atomic theory had become the province of the chemists. Beginning with Boyle the chemists had identified a variety of substances as "elements"—meaning that they could not be further reduced to other substances by chemical means. Several chemists concluded that all chemical reactions involved recombinations of these *elements*. Indeed, in 1799, J. L. Proust stated an empirical chemical law to the effect that in chemical reactions the elements involved combine in definite proportions, so that if there is too much, or too little, of one of the elements in a reaction, the surplus will be unable to combine with anything. In 1808 John Dalton, the great English chemist, published the first volume of *A New System of Chemical Philosophy*, in which these chemical regularities were explained by the assumption that each element consisted of characteristic atoms. A chemical reaction in this picture was simply the combination of these atoms, in various proportions, from among the different elements. Of the properties of the individual atoms—their size and shape and so on—nothing very specific could be said.

The next real progress came from a study of gases. In the first place the French chemist J. L. Gay-Lussac, in the beginning of the nineteenth century, found that gases react with each other in definite proportions by volume. (Two volumes of hydrogen combine with one volume of oxygen to produce two volumes of water.) This led the Italian physicist and mathematician Amedeo Avogadro to make the brilliant conjecture that under standard conditions of temperature and pressure *all gases* at a *given volume* contain the same number of particles. We refrain here from using the word "atom" because the gas particle in question may not be an atom, but rather a *molecule*—several atoms bonded together. (Ordinary oxygen gas, for example, consists of molecules composed of two oxygen atoms.) Avogadro wrote his basic paper in 1811, but it was not until 1858, when the Sicilian chemist Stanislao Cannizzaro published his *Sketch of a Course of Chemical Philosophy*, which used the molecular form of Avogadro's hypothesis to organize a variety of data on the chemistry of gases, that these ideas began to have some currency among chemists. At almost the same time James Clerk Maxwell and, independently and by different methods, Ludwig Boltzmann formulated the foundations of the statistical mechanics of such particle gases, thus generalizing the work begun by Bernouilli. The basic idea again was that the apparently orderly behavior of gases in the large can be deduced from the assumption of randomly occurring molecular events. In other words, they visualized a gas as consisting of molecules in constant random collision with each other. Each collision, if one could follow it, would be seen to be governed, in this classical picture, by the laws of Newtonian mechanics. But in practice one cannot follow each collision, since there are too many of them. Hence Maxwell and Boltzmann invented methods for determining the *average* behavior of such a large collection of molecules. In particular, they were able to predict that in a typical gas, at room temperature, molecules should be flying about with speeds of the order of a *thousand feet per second* if nothing got in the way. We know, however, from common experience, that it takes a substantial amount of time for a gas to "mix"—in, say, the air. The fact that this is a slow process although the molecules move very quickly is explained by the fact that a typical molecule suffers frequent collisions as it diffuses about. The number of these collisions depends on the size of the individual molecules and on the number of molecules that are available to get in the way.

In 1865, by studying the data on diffusion, J. Lodschmidt estimated that the then still hypothetical gas molecules would have diameters of about one fifty millionth of an inch (about a hundred millionth of a centimeter). Lodschmidt was also able to estimate the number of molecules in a given volume of gas under standard conditions of temperature and pressure. (These are taken, arbitrarily, to be zero degrees centigrade and one atmosphere of pressure.) The chemists choose this standard volume so that the gas contained in it weighs, in grams, the molecular weight of that gas—sixteen grams for, say, oxygen. The number of molecules in such a sample turns out to be, according to recent measurements, 6.0249×10^{23} molecules. This means that there are, under the same conditions, about 4.5×10^{19} molecules in a cubic centimeter of gas: *a lot of molecules*!

As clear as this statistical mechanical picture may appear to the contemporary scientist, it was, even at the end of the nineteenth century, an arcane and controversial one which was by no means widely accepted or even understood among physicists. Einstein, in his isolation in the patent office in Bern, reinvented the subject for himself. In the period from 1902 to 1904 he published three papers which consisted mainly of his rediscovery of work that had already been done by Boltzmann and by the extraordinary American physicist Josiah Willard Gibbs. Gibbs, whose work was done in the last quarter of the nineteenth century, was, if anything, even more isolated professionally than Einstein, and it was not until near the turn of the century that European physicists became aware of the full magnitude of Gibbs's contributions to statistical mechanics. Einstein later remarked that if he had known about the work of Gibbs and Boltzmann he would never have published his own early papers. But the fact that Einstein rediscovered this work made him an absolute master of the techniques of statistical mechanics. He had so much confidence in these ideas that he was able to apply them in all sorts of new domains, and, as we shall see, they were at the base of his principle contributions to quantum physics. In 1905 Einstein applied statistical mechanics to the

Brownian motion of the tiny objects suspended in liquids.[7]

The main strength of Einstein's approach to the theory of Brownian motion is its quantitative character. It is one thing to argue, as had been done, that the Brownian motion had something to do with molecular motion, but it is another thing to make *quantitative* predictions as to what will happen. To simplify his calculations Einstein represented Brown's particles as tiny spheres of radii something like a ten thousandth of a centimeter, while the typical molecule of the liquid, in which the much larger Brownian particles are suspended, has a radius of only a hundred millionth of a centimeter or so. These molecules collide constantly with the Brownian particles from all sides, and Einstein assumed that it was equally likely that in these collisions the Brownian particles got kicked in any possible given direction. Naïvely one might be tempted to think that a Brownian particle would simply be kicked back and forth and would never get anywhere in the liquid. But this is all but impossible, since once the Brownian particle is kicked away from its original position, subsequent collisions are more likely to take it farther away than to return it to where it started. What Einstein showed was that the average distance traveled by the Brownian particle increased as the *square root of the elapsed time,* so that given enough time the particle will move, or be moved, by the impacts of the molecules, arbitrarily far from its starting point.[8] The fact

---

[7] While the Brownian particles are "tiny," they are much larger than the molecules of the liquid in which they are suspended. The former can be seen with an ordinary microscope, while the latter cannot.

[8] For spherical Brownian particles Einstein's equation takes the form:

$$\Delta = k \sqrt{T t / r\eta}$$

In this equation Δ is the average displacement, t is the time, k is a constant that is universal for all liquids, T is the temperature of the liquid, r is the radius of the spherical

that it is the *square root* of the time that enters is a consequence of the random statistical character of the collisions. This square root is a very characteristic and essential feature of the Brownian motion and is the striking and novel aspect of Einstein's prediction. It means, for example, that in four seconds the Brownian particle will move not four times what it will move in one second but only *twice* as far. In particular, Einstein predicted that, at room temperature in water, a Brownian particle will diffuse an average distance of something like a ten thousandth of a centimeter in one second.

By 1908, the French experimental physicist J. B. Perrin had tested and confirmed Einstein's formula. Moreover, by actually observing the distance that the Brownian particles traveled he was able to deduce approximately the number of molecules per cubic centimeter in the liquid through which they were traveling. This number is in essential agreement with Lodschmidt's number for gases.

All of this convinced the remaining skeptics of the scientific validity of the atomic hypothesis. One of the most interesting cases is that of the Russian-born German physical chemist Friedrich Wilhelm Ostwald. Ostwald is considered one of the founders of modern physical chemistry, but he was of the belief that chemists should study only what is measurable—the transfer of energy in chemical reactions—and not make what he thought were vague hypotheses about things like atoms which were not directly observable. In this respect he

---

Brownian particle, and η is a number characteristic of each liquid which measures its "stickiness" or viscosity. The intuitive significance of these factors is clear. The greater the temperature the more agitated is the molecular motion and hence the more often and more violently will the Brownian particle be collided with. On the other hand the bigger the particle and the qluier the liquid, the less easily can it be moved by the collisions.

was a disciple of Mach. (Mach also disbelieved in atoms and was also converted by the Brownian motion.) However, by the end of the First World War, in a new edition of his *Outlines of Chemistry*, Ostwald was able to write:

> I am now convinced that we have recently become possessed of experimental evidence of the discrete or grained nature of matter for which the atomic hypothesis sought in vain for hundreds and thousands of years. The isolation and counting of gaseous ions on the one hand . . . and on the other the agreement of the Brownian movements with the requirements of the kinetic hypothesis . . . justify the most cautious scientist in now speaking of the experimental proof of the atomic theory of matter. The atomic hypothesis is thus raised to the position of a scientifically well-founded theory.[9]

A bizarre footnote to this situation has come to light recently.[10] Among the first letters written by Einstein, and still preserved, is one he wrote on March 19, 1901, to Ostwald asking for a job. Einstein had just been turned down for an assistantship at the Polytechnic in Zurich and, without a job, was inspired to write to Ostwald, not only because Ostwald was an outstanding scientist but because Ostwald held Machian positivistic views which were, at that time, close to Einstein's own. (In an earlier edition of his book, Ostwald had denied the existence of the ether.) However, there was no favorable response to Einstein's letter, nor to a second one written on April 3, nor to a third one written, unknown to Einstein, by his father on April 13 in which Hermann Einstein said that his son esteemed Ostwald

[9] Cited in Robert A. Millikan, *The Electron*, p. 10. Millikan, who made important contributions to the experimental study of Brownian motion, reviews the history of the subject in this book.
[10] See Gerald Holton, "Mach, Einstein and the Search for Reality," pp. 636–37.

"most highly among all scholars currently active in physics." Ostwald can hardly be blamed for this over-sight in 1901, since, at this time, Einstein's only publi-cation, on capillary action, gave no special indication of any outstanding future promise and since his academic record at the Polytechnic was hardly inspiring.

In reviewing Einstein's career, especially in the early years, one is continually struck by the difficulties his contemporaries had in fitting him into the normal spectrum of scientific achievement. By the time of the First World War every physicist realized Einstein was a creative genius, but his work was viewed at the time as so unconventional that it is difficult to find a physicist who had fully digested and appreciated the totality of it. This was due in part, perhaps, to the deceptive sim-plicity of the early papers—results seemed to materialize out of "thin air" rather than out of long chains of elab-orate calculations. Moreover, Einstein was continuously ahead of the experimental confirmations that even-tually established his predictions, or, perhaps more ac-curately, he was able, because of his intuition about gen-eral principles, to sort out experimental results he felt must be correct from the rest, which he more or less ignored

The attitude of the scientific establishment toward Einstein seems clearly illustrated when he eventually was awarded a Nobel Prize. From everything dis-cussed so far, the reader would certainly be entitled to the impression that if ever there were a physicist deserv-ing of the Nobel Prize, it was Einstein. By normal standards any one of his 1905 papers could have earned him a Nobel Prize. However, when Einstein *was* finally awarded the Nobel Prize, in 1922, it was for his work not on relativity or on Brownian motion, both of which were well-established branches of physics by then, but on the quantum, a subject at that time in a state of al-most total confusion. Almost everything about the details of this award—except the merits of its recipient —are exceedingly bizarre.

In the first place the committee did not announce its decision until November 10, 1922, although the award was for *1921*. Einstein did not actually collect the prize until April 1923, when the medal and diploma were delivered to him by the Swedish ambassador in Berlin. Even more peculiar was the wording of the citation: "ROYAL SWEDISH ACADEMY has at its assembly held on November 9, 1922, in accordance with the stipulation in the will and testament of Alfred Nobel dated November 27, 1895, decided to/ independent of the value that/ (after eventual confirmation) may be credited to the relativity and gravitation theory/ bestow the prize/ that for 1921 is awarded to the person within the field of physics who has made the most important discovery or invention/ to Albert Einstein being most highly deserving in the field of theoretical physics/ particularly his discovery of the law pertaining to the photoelectric effect."[11] The photoelectric effect was a subject that had occupied just *one section* of the first of the great 1905 papers—entitled "Concerning a Heuristic Point of View about the Creation and Transformation of Light."[12] It was as if the Swedish Academy was all but trying to rid itself of the ominous specter of the relativity theory.

The reason for this apparent caution had, no doubt, to do with the phrasing of the Nobel will. As is well known, Alfred Nobel, who had been trained as a chemist and engineer, and who had invented and patented dynamite and other relatively "safe" explosives from which he made a huge fortune, left most of his money, in trust, for annual prizes to be given to "those who have conferred the greatest benefit on man-

[11] I am grateful to Professor Holton for supplying the translation of the Swedish original.
[12] Translated and reprinted in Boorse and Motz, *op. cit.*, pp. 544–57.

kind" in one of the stated disciplines. This wording of
the will, at least as far as the physics prizes were concerned, acted to limit the subject matter of the eligible
discoveries. Most of the prizes, from the first award in
1901 until Einstein's, were for discoveries in experimental physics, and some of them, such as the 1912
award to Nils Gustaf Dalén for the "Invention of auto-
matic regulators for lighting coastal beacons and light
buoys during darkness or other periods of reduced visi-
bility" hardly seem to have much to do with science at
all. The committees have in general been cautious
about awarding prizes for speculative work unless it has
been confirmed by experiment beyond any doubt. Apparently, even as late as 1922, the Nobel committee felt
that the relativity theory was still too speculative to fall
under the aegis of the Nobel will.

As we have indicated, what is ironic about this is that
to the physicists of the period the quantum theory, of
which the photoelectric effect is an aspect, was surely
much more speculative than the relativity theory, and
much less well understood. To get some idea of the
confusion that existed one need only read a book like
The Electron, written in 1917 by the great American
experimental physicist Robert A. Millikan. (Millikan
was, in fact, awarded the Nobel prize for 1923, in part
for his work that confirmed Einstein's theory of the
photoelectric effect—much of which had been done a
decade earlier.) As late as 1917 Millikan could write:

> Despite then the apparently complete success of the
> Einstein equation [describing the photoelectric
> effect], the physical theory of which it was designed
> to be, the symbolic expression is found so untenable
> that Einstein himself, I believe, no longer holds to it
> [how Millikan got this idea is not explained, but, as
> far as I can tell, there is no basis for it; Einstein's
> rejection of "quantum theory" came after 1926 and
> had nothing to do with his work of 1905], and we
> are in the position of having built a very perfect

structure and then knocked out entirely the under-pinning without causing the building to fall. It stands complete and apparently well tested but without any visible means of support. These supports must obviously exist, and the most fascinating problem of modern physics is to find them. Experiment has outrun theory, or, better, guided by erroneous theory, it has discovered relationships which seem to be of the greatest interest and importance, but the reasons for them are as yet not at all understood.[13]

[13] Millikan, *op. cit.*, p. 236.

The Quantum

XV

The quantum—more precisely, the quantum of action—was invented and introduced into physics in 1900 by the German physicist Max Planck. It is somehow fitting that the theory of the quantum should have been born with the new century. Of all the ideas that we have considered up to now it is the only one to go, in a radically new way, beyond the bounds of classical physics. The concepts of the quantum theory really set twentieth-century physics off from anything that precedes it. This may seem like a strange observation after one has been exposed to the somewhat startling conclusions of the relativity theory. But the relativity theories, both special and general, are set in a philosophical context of a causal description of events occurring in space and time—or space-time. Einstein's space-time consists of points whose positions and times are to be determined by classical procedures using rulers and clocks of a kind that any nineteenth-century physicist would be comfortable with. Quantum theory, at least in the modern form developed in the 1920s, denies the underlying validity of such descriptions and, in so doing, has changed and influenced the whole epistemological basis of science.

As one might imagine, the full acceptance of these revolutionary ideas came slowly, and one finds again and again the curious spectacle of the progenitors making a dramatic step and then drawing back, standing aloof, while the next generation of scientists sweep on. Planck, for example, was one of the earliest of the great physicists to accept the special theory of relativity. He was at work on it almost immediately. It fitted into his philosophical outlook. But of Einstein's exploitation of the quanta—Planck's quanta no less—of the work that turned the quanta into physics, Planck could write, "[that] he may sometimes have missed the target in his speculations, as, for example, in his theory of light quanta, cannot really be held against him."[1] And Einstein himself, in 1926, after Werner Heisenberg had made the first breakthrough that led to the new quantum mechanics, wrote to Max Born, "Quantum mechanics demands serious attention. But an inner voice tells me that this is not the true Jacob. The theory accomplishes a lot, but it does not bring us close to the secrets of the Old One. In any case, I am convinced that He does not play dice."[2]

Recently an even more bizarre relationship of Ein-

[1] This phrase occurs in the proposal presented by Planck and several colleagues that Einstein be admitted to membership in the Royal Prussian Academy of Sciences in 1913. See, for example, J. Bernstein, *A Comprehensible World*, p. 120.

[2] The original letter is in German. The two published translations differ somewhat in style but not in essential content. That given here is from Martin J. Klein, "Max Born on His Vocation," *Science*, 169 (1970), 361.

stein to the development of the quantum theory has come to light. Of all of the aspects of the quantum theory the one that is most characteristic of its methodology is the Heisenberg "uncertainty principle," which gives the precise limitation to measurements on the atomic scale. The fact that there were such limitations imposed by the theory on the measurement of, for example, positions and momenta of particles, disturbed Einstein profoundly. It simply could not be reconciled with his view of physics in terms of the geometric structure of space-time. He made many unsuccessful attempts to refute the uncertainty relationships, although gradually he modified his position to maintaining that the relationships were "incomplete"—an intermediate step between classical physics and some ultimate synthesis involving classical concepts of fields and the geometry of space-time. The irony is that Heisenberg attributes his original inspiration in formulating the uncertainty principle to a conversation that he had with Einstein in 1926. Heisenberg still had the notion that Einstein then held the kind of Machian positivistic views—the idea that all quantities that entered a physical theory must have "operational definitions" in terms of measuring instruments which characterized the analysis leading to the special theory. He did not realize that Einstein had abandoned this position many years earlier when he was seeking his final formulation of the theory of gravitation. Hence Heisenberg was astounded when Einstein asked, "But you don't seriously believe that none but observable magnitudes must go into a physical theory?" To which Heisenberg answered, with some surprise, "Isn't that precisely what you have done with relativity? After all you did stress the fact that it is impermissible to speak of absolute time, simply because absolute time cannot be observed; that only clock readings, be it in the moving reference system or the system at rest, are relevant to the determination of time."

As Heisenberg recalls, Einstein replied, "Possibly I

did use this kind of reasoning but it is nonsense all the same. Perhaps I could put it more diplomatically by saying that it may be heuristically useful to keep in mind what one has actually observed. But on principle, it is quite wrong to try founding a theory on observable magnitudes alone. In reality the very opposite happens. It is the theory which decides what we can observe. . . ." This phrase—*"It is the theory which decides what we can observe"*[3]—remained in Heisenberg's mind and became the theme that led him to his invention of the uncertainty principle.

Now to return to Planck and the quantum. Planck himself was one of the least radical personalities imaginable. He was descended from an ancient German family of scholars, public servants, and lawyers. His father was a professor of law at Kiel, where Planck was born in 1858. "What led me," he wrote,

> to my science and from my youth filled me with enthusiasm is the fact—not at all self-evident—that our laws of thinking conform with the lawfulness in the passage of impressions which we receive from the outer world, thus making it possible for man to gain information about that lawfulness by mere thinking. [This idea was stated by Einstein in his famous aphorism, "One may say the eternal mystery of the world is its comprehensibility."] In this it is of the highest significance that the outer world represents something independent of us and absolute with which we are confronted, and the search for the laws which govern this absolute has appeared to me as the most fascinating work of a lifetime.[4]

Early in his scientific career Planck was struck by the fact that the great laws of thermodynamics involv-

[3] This conversation, as Heisenberg recalls it, can be found in his *Physics and Beyond: Encounters and Conversations*, p. 63.
[4] Cited in Max Born's very moving biographical study, "Max Karl Ernst Ludwig Planck," in Henry A. Boorse and Lloyd Motz, eds., *The World of the Atom*, p. 463.

ing energy and entropy seem to have just this absolute character. For this reason he at first rejected the idea of Maxwell, and especially Boltzmann, that the laws of thermodynamics really summarized only the *average* or most *probable* behavior of systems with enormous numbers of particles. For Boltzmann, who had seen the deepest into these ideas by the mid-1880s, a system like a gas evolved in time toward configurations of greater and greater probability. During the course of the evolution the system might "fluctuate" into a configuration that was improbable, but then it would continue its evolution toward the probable. For Planck, at first, such probability considerations violated his ideal of the "absolute," and so he rejected them. However, by the turn of the century, under pressure from new experimental results, he was forced to change his mind.

The events that led to Planck's quanta began rather innocently. Several physicists, notably Gustav Kirchhoff, who was Planck's teacher at the University of Berlin, began in the mid-nineteenth century a theoretical and experimental attack on the problem of so-called "black body" radiation. In a typical experimental setup one heats a partially evacuated enclosure to some temperature. As we now picture what happens, the atoms that make up the material of the walls of the enclosure begin to vibrate as the walls are heated, and these vibrators—or oscillators—give off electromagnetic radiation. This radiation bounces around the walls of the enclosure and eventually an equilibrium situation is established. The radiation inside the enclosure has various frequencies—colors, if you will—and these are distributed in a certain spectrum which physicists call the "normal" or "black body" spectrum. This distribution can be measured by making a small hole in the enclosure and sampling the radiation inside with a spectrometer and, say, a thermocouple to measure the intensity at a given frequency. What Kirchhoff proved was that this spectrum was independent of the nature of the material making up the walls. Hence it had the sort of "absolute" character that appealed to Planck. Moreover in the late 1890s the German physicist Wilhelm Wien found a very simple mathematical expression—he guessed at it—that appeared to fit the experimental results on the form of the spectrum and Planck "eagerly set to work" trying to derive from the laws of thermodynamics the Wien spectrum.[5] By the end of the century Planck thought, erroneously, that he had found an argument for it, and in 1899 he had submitted a paper to the *Annalen der Physik* which he thought settled the question. However, while he was correcting the proofs of his paper, some new experimental results began to emerge, and in a note added in proof he pointed out that the Wien distribution did not seem to hold, especially at the end of the spectrum corresponding to the longer—that is, the redder—wave lengths.

By October 1900 Planck had realized that his original arguments for the Wien law were wrong and had invented a new formula—again with no real theoretical justification—which seemed to fit all of the experiments. In fact, as Planck recalled, no sooner had he announced his new formula in a meeting of physicists, than it was confirmed.

> The very next morning I received a visit from my colleague Rubens. He came to tell me that after the conclusion of the meeting, he had that very night checked my formula against the results of his measurements and found satisfactory concordance at every point. . . . Later measurements too, confirmed my radiation formula again and again—the finer the methods of measurement used, the more accurate the formula was found to be.[6]

And so it has remained to the present day.

[5] See Martin J. Klein, "Max Planck and the Beginnings of the Quantum Theory," *Archive for History of Exact Sciences*, I, No. 5 (1962), 460.
[6] Quoted in *ibid.*, p. 465.

None of this had, as yet, to do with the quanta. These emerged at the next step, when Planck attempted to *derive* his formula. In this derivation he found reluctantly that he was forced to adopt Boltzmann's statistical methods—but with an important new assumption. To get his formula he was forced to assume that the atomic oscillators that constituted the walls of the radiation container could change energy, when they emitted and absorbed the radiation, from the cavity only in quantized units. In the description of this process according to classical physics, the oscillators could take up any amount of energy, but what Planck had to assume was that each oscillator could absorb units of energy that were integer multiples of some minimum unit characteristic of the oscillator. It could absorb one unit or two units and so on, but never half a unit or three quarters of a unit. The image that Einstein was fond of was that, like beer in a keg which can be bought and sold only in pint bottles, so the radiation energy could be taken from the atomic oscillators only in discrete, lump sums—quanta of energy.

Planck was at first quite confident that this represented some flaw in his method of derivation, and that if only he worked a little harder he could make the quanta go away and still obtain his formula. He spent the next ten years trying unsuccessfully to do this. As he wrote late in his life—Planck died in 1947—

> My futile attempts to fit the elementary quantum of action somehow into the classical theory continued for a number of years, and they cost me a great deal of effort. Many of my colleagues saw in this something bordering on a tragedy. But I feel differently about it. For the thorough enlightenment I thus received was all the more valuable. I now knew for a fact that the elementary quantum of action played a far more significant part in physics than I had originally been inclined to suspect.[7]

[7] Max Planck, *Scientific Autobiography and Other Papers*, trans. F. Gaynor (New York, 1949), p. 44.

Einstein's starting point was much less ambitious and much less radical, although, as usual, by following his arguments through to the end no matter where they led, he arrived at conclusions that altered our view of the physical world forever. Einstein began by accepting the Wien law—essentially identical to Planck's formula for the shorter, or more violet, wave lengths—as an experimentally given fact; and he asked himself what this implied about the radiation in the cavity. (It was only several years later that he tried to derive the full Planck formula—a derivation that is now standard in all the textbooks on quantum physics.) By using his knowledge of statistical mechanics he was able to show from the Wien law that the radiation *in* the enclosure obeyed the same mathematics as if it were made up of quanta. In other words, in Einstein's image, it was not only that the beer was bought and sold in pints from the keg, but that the beer in the keg consisted *only* of pints. There was no escape from this conclusion so long as one accepted the Wien law—or equivalently, for the short wave lengths, the Planck law—from experiment and applied the same reasoning that was totally successful for the statistical mechanics of gases and liquids, the same reasoning that Einstein had used in his theory of Brownian motion. Having established this, he then asked the next question—which Planck had not done—which was whether this discovery had any further implications for other phenomena involving the interaction of radiation and matter. He found several examples, and among them was the "photoelectric" effect.

The photoelectric effect first came to the attention of physicists in the late nineteenth century. It was noticed that if ultraviolet—very high frequency—light fell on the surfaces of certain metals, it could cause, from time to time, the ejection from the metal of a very high-energy electron. What was peculiar about this was that, since it was assumed, especially since the days of Maxwell, that electromagnetic radiation consisted of *waves*, one could not understand why, if a wave had enough energy to knock out a single electron, it did not knock out *all* the electrons in its path. It was as if a group of swimmers were lined up on a beach in the path of a wave that was apparently energetic enough to knock them all over, and, instead of doing so, it simply, every once in a while, knocked down a single swimmer, leaving the rest intact. In 1902 Philipp Lenard, the experimental physicist who won the Nobel Prize in 1905 and later became a virulent Nazi and a leading spokesman against "Jewish physics," made the important observation that if the intensity of the light was increased— by, for example, bringing the light source closer to the metal plate—the energy of the ejected electrons did *not* increase but, rather, only their number. This was also totally mysterious from the wave point of view. In the swimmer analogy it would be as if when one increased the amplitude of the wave the effect was not to knock the swimmers down harder, but rather to knock more of them down with exactly the same force that the smaller wave had done.

It was this set of facts that Einstein wanted to explain with the quanta. Here he made use of an important consideration that had emerged from his calculations. He had shown—and this was also a feature of Planck's derivation of the black-body spectrum—that the energy of a light quantum was proportional to its frequency. This meant, for example, that quanta of blue light would be more energetic than quanta of red light, and that X-rays would be more energetic still since their frequencies were larger than any visible light frequency. The basic energy equation, then, takes the simple form $E = h\nu$ where E is the energy of the quantum, ν is the frequency, and h is a new natural constant which is now called Planck's constant. (If E is measured in ergs and the frequency is measured in seconds then Planck's constant has, according to recent measurements, the value 6.62559×10^{-27} erg seconds, with a small experimental error in the last two places. This is a "small" number, which explains why quantum phenomena are too small to be observed in everyday life.)

If one accepted the quantum hypothesis one could explain Lenard's result in one fell swoop. According to this picture the light acts here not like a wave but rather as a collection of discrete energy quanta. So long as the light is of a single frequency—say blue—all the quanta will have the same energy. No matter how intense the source is it can only produce more quanta of the same energy. When a light quantum collides with an electron it can give up all of its energy to the electron—like billiard balls in collision—and the more intense the source the more quanta there are and hence the more collisions can be produced.

While Lenard's experiments were in qualitative agreement with these ideas, the first quantitative experiments were not done until a decade later by Millikan, who published his results in 1916. Millikan was an extraordinarily gifted experimenter, and in making these experiments he had to maintain his metal surfaces at high vacuum to keep from having them coated with impurities that would block the exit of the electrons from the metal. He constructed a sort of miniature "barber shop" inside an evacuated glass container with a remotely controlled "razor" which he could use to shave the metal surfaces. By varying the color of the impinging light over a rather wide range he was able to make a full and successful test of Einstein's equation.

As we have seen, this was not a fact which Millikan greeted with great elation, because it seemed to him, and to most other physicists, including Einstein, that this merely deepened the mystery as to the nature of light. Until the beginning of the nineteenth century it had been assumed by most physicists that light *did* consist of energetic particles (a notable exception was the seventeenth-century physicist Christian Huygens, who made the first important steps toward a wave theory of

light; his work was ignored, under the influence of Newton, who favored the light particles), but new experiments, begun in England by Thomas Young and followed up in France by Augustin Fresnel and others,
seemed to rule out particle theories of light once and
for all. All these experiments have a common theme—
namely, "interference." When two waves overlap in
space they interfere with each other, which means that
they combine to produce a wave motion that differs
from each of the individual wave motions. In fact at
certain points in the resultant wave train—if a trough
of one wave is superposed on a crest of the other—the
resultant wave amplitudes can exactly cancel, giving a
null amplitude; two light waves can interfere so as to
produce a dark spot. This Young tested by passing light
through two small holes in a screen and noticing that
the combined effect of the light coming through the two
holes was to produce a pattern of alternating bands of
light and dark arising from the constructive and destruc-
tive interference of the two trains of light waves. In fact
Young was able to use this technique to measure the
wave length of various kinds of visible light.

During the nineteenth century these techniques were
refined, and the whole science of optics was built around
the wave theory of light, which appeared to have found
its final formulation in Maxwell's equations. Needless
to say, Einstein was fully aware of the success of the
wave theory when he wrote his 1905 paper on light
quanta. Indeed, he begins his paper by stating flatly,
"The wave theory, operating with continuous spatial
functions, have proved to be correct in representing
purely optical phenomena and will probably not be replaced by any other theory."

But [he goes on] one must, however, keep in mind
that the optical observations are concerned with tem-
poral mean values and not with instantaneous values,
and it is possible, in spite of the complete experimental verification of the theory of diffraction, reflec-

tion, refraction, dispersion, and so on, that the theory of light that operates with continuous spatial functions may lead to contradictions with observations if we apply it to the phenomena of the generation and transformation of light.

It appears to me, in fact, that the observations on "black body radiation," "photoluminescence," the generating of cathode rays [electrons] with ultraviolet radiation, and other groups of phenomena related to the generation and transformation of light can be understood better on the assumption that the energy in light is distributed discontinuously in space. According to the presently proposed assumption the energy in a beam of light emanating from a point source is not distributed continuously over larger and larger volumes of space but consists of a finite number of energy quanta, localized at points of space, which move without subdividing and which are absorbed and emitted only as units.[8]

For the next twenty years Einstein spent much of his time trying to understand this "schizophrenic" character of light. (The reader may have noted that the schizophrenia is already built into the fundamental energy equation of the quanta $E = h_\nu$, since the frequency ν is a wave concept but also characterizes the quantum "particle" energy.) This work seems to have divided itself into two parts, a part that Einstein published and that had to do with finding new applications of the quantum ideas to other phenomena, and a part that he did not publish, because it was unsuccessful, that had to do with finding some sort of basic theory which incorporated these two aspects of light in a single description. Of the former, the two most interesting parts are perhaps Einstein's quantum theory of the "specific heats" of solids and his study of that part of the black-body spectrum that corresponded to *long* wave lengths. Specific heat refers to the ability of objects to absorb

[8] Henry A. Boorse and Lloyd Motz, eds., *The World of the Atom*, pp. 544–45.

heat and thereby raise their temperature; the amount by which the temperature of a body is raised when a specific amount of heat is added. Einstein found that some anomalies in the observed behavior of such heat absorption at low temperatures could be explained by using Planck's *quantized* oscillators as the heat absorbers instead of the classical absorbers which could take up arbitrary amounts of energy. This founded the modern development of the quantum theory of solids, and since there were confirming experiments available in 1907, when Einstein did his first work on the subject, it attracted the attention of many physicists to the quantum hypothesis for the first time.

The second important piece of work had to do with the long wave-length part of the Planck formula. This is the other end of the spectrum from that which Einstein had considered in his original 1905 paper. In fact— and Planck had not appreciated this at the time he was doing his work—a theory that fitted this end of the spectrum had been worked out by Lord Rayleigh, and later slightly modified by the British astronomer James Jeans, in 1900. The important feature of their result is that it is an inevitable consequence of classical physics. In fact if Planck's constant, which characterizes the quanta, is set equal to zero in his formula, one arrives at the Rayleigh-Jeans law. Einstein went one step further. He showed, again using statistical mechanical arguments, that this end of the spectrum arises from the *wave* nature of light. Hence, the full black-body spectrum has built into its structure both the wave and particle aspects of light in a striking fashion.

Some insights into Einstein's state of mind as he wrestled with the quanta can be gleaned from an anecdote and a letter. The anecdote has been recounted by Philipp Frank:

About this time [Einstein was then in Prague] . . . Einstein began to be much troubled over the paradoxes arising from the dual nature of light. . . . His state of mind over this problem can be described by this incident:

Einstein's office at the university overlooked a park with beautiful gardens and shady trees. He noticed that there were only women walking about in the morning and men in the afternoon, and that some walked alone sunk in deep meditation and others gathered in groups and engaged in vehement discussions. On inquiring what this strange garden was, he learned that it was a park belonging to an insane asylum of the province of Bohemia. The people walking in the garden were inmates of this institution, harmless patients who did not have to be confined. When I first went to Prague, Einstein showed me this view, explained to me, and said playfully: "Those are the madmen who do not occupy themselves with the quantum theory."[9]

The letter was written somewhat earlier, in 1908, to a collaborator named J. J. Laub:

I am ceaselessly occupied with the constitution of radiation and am in correspondence on this question with H. A. Lorentz and Planck. The former is astonishingly profound [Lorentz, then fifty-five, was regarded as the greatest living theoretical physicist, while Einstein was twenty-nine and still working at the patent office] and at the same time, amiable man. Planck is also very pleasant in his correspondence. He has, however, one fault: that he is clumsy in finding his way about in foreign trains of thought. It is therefore understandable when he makes quite faulty objections to my latest work on radiation. He has not, however, said anything against my criticisms. I hope that he has read them and recognized them. This quantum question is so incredibly important and difficult that everyone should busy himself on it. I have already succeeded in working out something

[9] Philipp Frank, *Einstein: His Life and Times*, p. 98.

which may be related to it but I have serious reasons for still thinking that it is rubbish.[10]

Einstein had moved from the University of Zurich to Prague in 1911, but in the autumn of 1912 he was back at Zurich, this time as a professor at the Polytechnic from which he had graduated, without special distinction, over a decade earlier. He remained in Zurich for less than two years and in April 1914 moved to Berlin, where he remained until he left Germany forever, in 1932.

During this period Max Born was in Berlin frequently; nominally he held a chair at the University of Berlin, but most of the normal activities at the university were suspended after the military mobilization in August 1914. Nonetheless, with the presence of Einstein, Planck, and, a little later, Erwin Schrödinger, Berlin became and remained until the mid-1920s (when the center of gravity shifted to younger men in Göttingen and Copenhagen) the greatest center in the world for theoretical physics—to say nothing of the great experimental physicists and chemists who were there. The Berlin physics colloquium had a regular weekly attendance of outstanding scientists the like of which has never been assembled for any length of time in one place before or since.

Planck and Einstein met at regular intervals at the Berlin Academy, and a friendship developed which went far beyond the exchange of scientific ideas. Yet it is difficult to imagine two men of more different attitudes to life: Einstein a citizen of the whole world, little attached to the people around him, independent of the emotional background of the society in which he lived—Planck deeply rooted in the traditions of his family and nation, an ardent patriot, proud of the greatness of German history and consciously Prussian in his attitude to the state. Yet what did all these

[10] Quoted in Martin J. Klein, "Thermodynamics in Einstein's Thought," 513.

differences matter in view of what they had in common—the fascinating interest in the secrets of nature,
similar philosophical convictions and a deep love of
music. They often played chamber music together,
Planck at the piano and Einstein fiddling, both perfectly absorbed and happy. Planck was an excellent
pianist and could play on demand almost any piece of
classical music, a great many by heart. He also liked
to improvise either on a theme given to him, or on
old German folk-songs which he dearly loved.[11]

Einstein was sensitive to the dangers of the rise of
Nazism in Germany from its beginning in the early
1920s when the first outbreaks of organized German
anti-Semitism began. Planck, on the other hand, because perhaps of his sense of patriotism or his advanced
age, underestimated the strength and durability of the
Nazi movement. He despised the Nazis but even then
was convinced that they represented only a passing
phase. At one point, around 1933, he had an audience
with Hitler in order to try to spare the career of a Jewish colleague, the celebrated chemist Fritz Haber.
(Haber had made discoveries concerning the manufacture of explosives during the First World War which,
according to Born, were crucial in keeping Germany in
the war at all.) Hitler was transported into such a rage
that, in the words of Born, "Planck could do nothing
but listen silently and take his leave."[12] After this,
Planck abandoned hope of actively changing the regime
and resigned himself to living out his existence in Germany as best he could. His house and library were destroyed by air raids during the Second World War and
his son, Erwin, was executed by the Nazis after he had
taken part in the July 1944 plot against Hitler. Planck
lived the remainder of his life in Göttingen, where he
died in 1947 at the age of eighty-eight.

[11] Max Born's obituary for Planck, in Boorse and Motz, *op.
cit.*, p. 477.
[12] *Ibid.*, p. 483.

The Middle Years

XVi

The post-First World War years were hard ones
for Einstein. He was frequently ill. The German
inflation had made his financial position very
difficult, especially since he was supporting his
former wife and two sons in Switzerland, and
he found the general atmosphere of Germany
menacing and uncongenial. In fact, had it not
been for his loyalty to Planck and a few other
colleagues, he might well have migrated to Holland to join his good friend Paul Ehrenfest in
Leyden. Ehrenfest, born in Vienna in 1880, had
first met Einstein in Prague in 1912 and had
been so impressed with him that he had been
prepared to move to Zurich just to have the opportunity to work with him. Einstein found in
Ehrenfest a wonderfully enthusiastic and totally
honest and rigorous scientific critic, and the
two men developed a deep friendship—at least

as deep a friendship as Einstein's nature allowed him to form. And after Ehrenfest went to Leyden in 1912 as Lorentz's successor, Einstein visited him fairly frequently. One visit took place in 1916, just after Einstein had put the general theory of relativity in its final form. In addition to Ehrenfest, Einstein wanted very much to talk to Lorentz, who although in nominal retirement in his early sixties, was, in fact, still working intensely in research. Ehrenfest wrote a vivid account of this meeting with Lorentz:

> In his usual way Lorentz saw to it first at dinner that Einstein felt himself enveloped in a warm and cheerfull atmosphere of human sympathy. Later, without any hurry, we went up to Lorentz's cozy and simple study. The best easy chair was carefully pushed in place next to the large work table for the esteemed guest. Calmly, and to forestall any impatience, a cigar was provided for the guest, and only then did Lorentz begin quietly to formulate a finely polished question concerning Einstein's theory of the bending of light in a gravitational field. Einstein listened to the exposition, sitting comfortably in the easy chair and smoking, nodding happily, taking pleasure in the masterly way Lorentz had rediscovered, by studying his works, all the enormous difficulties that Einstein had to overcome before he could lead his readers to their destination, as he did in his papers, by a more direct and less troublesome route. But as Lorentz spoke on and on, Einstein began to puff less frequently on his cigar, and he sat up straighter and more intently in his armchair. And when Lorentz had finished, Einstein sat bent over the slip of paper on which Lorentz had written mathematical formulas to accompany his words as he spoke. The cigar was out, and Einstein pensively twisted his finger in a lock of hair over his right ear. Lorentz, however, sat smiling at an Einstein completely lost in meditation, exactly the way a father looks at a particularly beloved son—full of secure confidence that the youngster will crack the nut he has given him, but eager to see how. It took

quite a while but suddenly Einstein's head shot up
joyfully; he "had it!" Still a bit of give and take, inter-
rupting one another; a partial disagreement, very
quick clarification and a complete mutual understanding, and then both men with beaming eyes
skimming over the shining riches of the new theory.[1]

It is little wonder that toward the end of his own life
Einstein could write of Lorentz, "For me personally he
meant more than all the others I have met on my life's
journey."[2] And it is little wonder that Einstein was
tempted to settle in Leyden in 1919 after Ehrenfest
communicated an offer which would have allowed Einstein both financial security and a complete option to
spend his time as wished "provided only that one can
say 'Einstein is in Leyden—in Leyden is Einstein.'" A
few days later Einstein wrote:

> Your offer is so fabulous and your words are so
> friendly and so full of affection that you can hardly
> imagine how confused I have been as a result of your
> letter. You know, of course, how delighted I am with
> Leyden. And you know how much I like all of you.
> But my situation is not so simple that I can do the
> right thing by just following the inclination of my
> own feelings. I am sending you a letter that Planck
> wrote me while I was in Zurich. After receiving it I
> promised him not to turn my back on Berlin unless
> conditions were such that he would regard such a
> step as natural and proper. You have hardly any idea
> of the sacrifices that have been made here, with the
> general financial situation so difficult, to make it pos-
> sible for me to stay and also to support my family in
> Zurich. It would be doubly base of me if, just when
> my political hopes are being realized, I were to walk
> out unnecessarily, and perhaps in *part* for my ma-
> terial advantage, on the very people who have sur-
> rounded me with love and friendship, and to whom
> my departure would be doubly painful at this time of

[1] Cited in Martin J. Klein, *Paul Ehrenfest*, pp. 303–304.
[2] Quoted in *ibid.*, p. 303.

supposed humiliation. You have no idea with what
affection I am surrounded here; they are not all out to
catch the drops of oil my brain sweats out.[3]

And so Einstein remained in Berlin, although he did
accept a special professorship in Leyden that enabled
him to visit for a few weeks each year.

In the same letter to Ehrenfest Einstein asks, "Have
you perhaps heard of anything there about the English
solar eclipse expedition?" It was this expedition, as we
have seen, that first confirmed the bending of light by
the sun. Einstein's letter was written on September 12,
1919, and on the twenty-seventh he had received
word from Lorentz that his theory had been confirmed.
This made Einstein a public figure overnight; and, perhaps for that reason, and the fact that he was Jewish,
this coincided with the first well-organized political and
anti-Semitic attacks against the relativity theory. Several
decades later, this whole horrible episode is beginning
to assume the proportions of a terrible dream that we
can, only with difficulty, associate with reality. For the
people who lived through it, especially for someone like
Einstein who was both a Jew and a humanitarian, it left
scars that never healed. In 1946 when the German
physicist Arnold Sommerfeld, who had been an anti-
Nazi but who had, like Planck, remained in Germany,
wrote to Einstein to suggest that he might be interested
in renewing his membership in the Bavarian Academy
from which Einstein had been expelled in 1933, Einstein replied, "The Germans slaughtered my Jewish
brethren; I will have nothing further to do with them,
not even with a relatively harmless academy. I feel
differently about the few people who, insofar as it was
possible, remained steadfast against Nazism. I am happy
to learn you were among them."[4]

[3] This correspondence is given in full in *ibid.*, pp. 310–12.
[4] Quoted in Otto Nathan and Heinz Norden, eds., *Einstein
on Peace*, pp. 367–68.

In 1920 an anti-Einstein League was formed in Germany, and it offered substantial sums of money to anyone who would write refutations of Einstein's work. On August 24, 1920, the League sponsored a meeting in the Berlin Philharmonic Hall, which Einstein himself attended, where swastikas and anti-Semitic pamphlets were on sale, at which Einstein and his work were attacked. A few of his colleagues responded in a letter to the *Berliner Tageblatt* and a few days later Einstein himself wrote an angry letter, also published in the *Berliner Tageblatt*—which deeply shocked Ehrenfest, who seemed to feel that Einstein should have ignored the matter as unworthy of his attention. From this time on, until he finally left Germany in 1932, Einstein and his work were the targets of a steadily mounting campaign.

To read even a very brief account of the fate of German Jews in the 1930s is to make one sick at heart. By 1933 Philipp Lenard could write in the *Völkische Beobachter*, the Nazi paper:

> The most important example of the dangerous influence of Jewish circles on the study of nature has been provided by Herr Einstein with his mathematically botched-up theories consisting of some ancient knowledge and a few arbitrary additions. This theory now gradually falls to pieces, as is the fate of all products that are estranged from nature. Even scientists who have otherwise done solid work cannot escape the reproach that they allowed the relativity theory to get a foothold in Germany, because they did not see, or did not want to see, how wrong it is, outside the field of science also, to regard this Jew as a good German.[5]

Two years later Lenard delivered an inaugural address at the opening of a new physics institute:

[5] Quoted in Philipp Frank, *Einstein: His Life and Times,* p. 232.

I hope that the institute may stand as a battle flag against the Asiatic spirit in science. Our Führer has eliminated this same spirit in politics and national economy, where it is known as Marxism. In natural science, however, with the overemphasis on Einstein, it still holds sway. We must recognise that it is unworthy of a German to be the intellectual follower of a Jew. Natural science, properly so called, is of completely Aryan origin and Germans must today also find their way out into the unknown. *Heil Hitler.*[6]

By 1939, despite statements like that of the Nazi Minister of Education Bernhard Rust that "National Socialism is not an enemy of science, but only of theories"[7]— science in Germany had been destroyed. It is only now recovering. Most of the great German scientists, Jewish or non-Jewish, had left. A small number of the older generation like Max von Laue managed to survive despite an outspoken anti-Nazi position, and some, like Heisenberg and von Weizsäcker, while never members of the Nazi Party made whatever compromises were necessary in order to continue to work. Von Weizsäcker, an outstanding nuclear physicist, was the son of Ribbentrop's Secretary of State, and as a member of a family of diplomats he tried to work out various arrangements with the Party so that the results of the relativity theory could be taught and used in research by divorcing them from their connection to Einstein and "Jewish physics." Once, in 1943, von Laue was rebuked for having given a lecture in Sweden in which the relativity theory was mentioned without adding that German physicists "expressly dissociated" themselves from it. "Von Weizsäcker advised the great physicist to reply that the theory had in fact been largely developed by the Aryans Lorentz and Poincaré long before Einstein. Von Laue ignored the friendly advice and sent an openly defiant

6 Quoted in *ibid.*, p. 232.
7 Quoted in *ibid.*, p. 233.

article to a scientific periodical about the theory: That shall be my answer, he wrote to von Weizsäcker."[8]

During this time Einstein's attitude toward many political and moral questions changed. Despite his deeply pacificistic feelings he urged European countries to begin to rearm. In 1933 he wrote to a young pacifist:

> What I shall tell you will greatly surprise you. Until quite recently we in Europe could assume that personal war resistance constituted an effective attack on militarism. Today we face an altogether different situation. In the heart of Europe lies a power, Germany, that is obviously pushing towards war with all available means. . . . Imagine Belgium occupied by present-day Germany! Things would be far worse than in 1914, and they were bad enough even then. Hence I must tell you candidly: were I a Belgian, I should not, in the present circumstances, refuse military service; rather, I should enter such service cheerfully in the belief that I would thereby be helping to save European civilization. . . .
>
> This does not mean that I am surrendering the principle for which I have stood heretofore. I have no greater hope than that the time may not be far off when refusal of military service will once again be an effective method of serving the cause of human progress.[9]

The more Einstein became aware of German anti-Semitism the closer a bond he felt to his fellow Jews. There is no more moving photograph of Einstein anywhere than one taken in a Berlin synagogue in 1930. There he sits—skeptic and free thinker that he was and remained until the end of his life—his unruly hair flowing from underneath the traditional black *yamulke*, holding his violin prepared to play in a concert for the purpose of raising money to help his fellow Jews. In the

[8] David Irving, *The German Atomic Bomb* (New York, 1967), p. 177.
[9] *Einstein on Peace*, p. 229.

background one can make out the congregation and can only imagine, with grief, what was to be their fate. Despite his strongly antinationalistic feelings Einstein became in the 1920s a public supporter of Zionism, as he saw in the movement a way of survival and hope for the European Jews. After the death of Chaim Weizmann in 1952, Einstein received an offer to become Israel's second president. He replied:

> I am deeply moved by the offer from our state of Israel, and at once saddened and ashamed that I cannot accept it. All my life I have dealt with objective matters, hence I lack both the natural aptitude and the experience to deal properly with people and to exercise official functions. For these reasons alone I should be unsuited to fulfill the duties of that high office, even if advancing age was not making increasing inroads on my strength.
>
> I am the more distressed over these circumstances because my relationship to the Jewish people has become my strongest human bond, ever since I became fully aware of our precarious situation among the nations of the world.[10]

---

[10] *Ibid.*, pp. 572–73.

God's Dice

XVii

Until the mid-1920s Einstein made fundamental
positive contributions to the quantum theory,
although, after 1925, when the theory seemed to
have made a decisive step forward at the hands
of Heisenberg, Pauli, Born, Bohr, Dirac, Schrödinger, and others, he turned against it. Perhaps
some future generation of physicists will, somehow, discover that Einstein's critical intuitions
were really right, though now this does not seem
likely. In fact, because of his opposition to the
later developments in the quantum theory, many
physicists have not been aware of the full scope
of Einstein's contributions to the quantum
physics of the 1920s. It is only recently that
historians of modern science have begun to
untangle the complicated web of influences that
led to the invention of the quantum theory and
the discovery of the wave nature of matter—only
to find, once again, Einstein as a pivotal figure.

Einstein's great work of the 1920s had to do with the quantum statistical mechanics of gases—what we now call the Bose-Einstein gas. S. N. Bose, an Indian physicist working in Dacca, had sent Einstein a short manuscript in English in 1924, and in July of that year, the *Zeitschrift für Physik* received a rather unusual communication: a paper written by Bose, entitled "Planck's Law and the Hypothesis of Light Quanta," but submitted by Einstein.[1] After reading the English manuscript Einstein had been so impressed by it that he had translated it himself into German and had sent it, on Bose's behalf, to the journal.

Bose had discovered a new method of doing quantum statistics and had applied it to give a new derivation of Planck's radiation formula. Einstein realized that the same methods could be applied to ordinary gases but that if one did so it would necessarily imply that the "particle" composing such gases would behave statistically like light quanta and therefore should, or *could*— since there was no experimental evidence at that time— also show wave characteristics. It was while he was carrying out these calculations that he received a copy of de Broglie's thesis in which de Broglie—for other reasons—had made the same conjecture and, in Einstein's words, had "lifted a corner of the great veil."[2] By 1925, Einstein had sufficient confidence in the whole scheme so that he could write, "A beam of gas molecules which passes through an aperture must, then, undergo a diffraction analogous to that of a light ray."[3] That this in fact happens was confirmed by C. J. Davisson and L. H. Germer and, independently, by G. P. Thomson in their celebrated experiments of 1927. (These experiments actually used electrons, but the principle is the same.) This means that not only is light schizophrenic

[1] See Martin J. Klein, "Einstein and the Wave-Particle Duality," p. 26.
[2] Quoted in *ibid.*, p. 26.
[3] Quoted in *ibid.*, p. 4.

—exhibiting both wave and particle aspects—but "particles" such as the electron are equally schizophrenic. Under suitable circumstances electrons act like "waves," i.e., beams of electrons can interfere with each other like interfering beams of light waves.

In 1925 Erwin Schrödinger and Heisenberg made the great theoretical breakthroughs that led to modern quantum mechanics. At first it looked as if each of them had invented a different theory, so that there were two separate quantum theories. But by 1926 Schrödinger proved that the two theories were mathematically equivalent. He discovered the equation that described the behavior of the de Broglie waves or, as he insisted on calling them, "the de Broglie–Einstein" waves. As he wrote, "My theory was stimulated by de Broglie's thesis and by short but infinitely far-seeing remarks by Einstein."[4] Einstein greeted Schrödinger's first paper with enthusiasm, and he wrote to Schrödinger, "The idea of your article shows real genius."[5]

Both Einstein and Schrödinger rejected the next step, which was taken by Born and his colleague Pascual Jordan. This had to do with the question of what the waves were supposed to represent. The first interpretation, going back to de Broglie, was that particles such as the electron were classical particles—small billiard balls—but that the waves determined the trajectories of these particles when, for example, they circulated around an atomic nucleus. This was a picture that was based as closely as possible on classical physics. But Born and Jordan argued that it was inconsistent and that the only possible interpretation of the waves was that from them—from their mathematical form as given

[4] Quoted in *ibid.*, p. 4.
[5] Much of the correspondence between Einstein and Schrödinger on the quantum theory has been reprinted in translation by Martin J. Klein in *Letters on Wave Mechanics* (New York, 1967). This remark was added on a postscript to a letter Einstein wrote to Schrödinger on April 16, 1926.

by the Schrödinger equation—one could calculate the *probable* behavior of the particles and *nothing else*.

One may illustrate this graphically with the light quanta. As we have seen, if a light ray is sent through a small aperture, because of the interference effects, a diffraction pattern will be formed on a screen at the other side of the aperture. Now the same experiment can be performed by sending light quanta, one at a time, through the aperture. Each light quantum will pass through and hit some point on the screen. According to the Born-Jordan interpretation one cannot predict with certainty what an individual light quantum will do when it arrives at the aperture. One can only state what it is *most likely* to do—something that is determined by the Schrödinger wave function; i.e., the solution to the Schrödinger wave equation.

In fact, according to the theory, the light quantum is most likely to hit that point on the screen where, according to the wave picture, the diffraction pattern is brightest. This resolves the wave-particle schizophrenia, but at the cost of giving up a deterministic description of physical events. Strict determinism in the sense that physicists had become accustomed to from Newton to Einstein had to be abandoned, and this Einstein, and Schrödinger as well, would not accept. As Einstein said again and again, "God does not play dice with the world."

The next steps were taken by Heisenberg and Bohr at Bohr's institute in Copenhagen, where Heisenberg was a frequent visitor. The main consequence of this work was the idea that the particle-wave duality was not an incidental feature of atomic physics but was, rather, a basic fact about nature that could be traced back to a careful analysis of the meaning of "measurement" on the atomic scale, which Heisenberg expressed in terms of his "uncertainty principle." The most famous elementary example is the "Heisenberg microscope." This is an imaginary device capable of generat-

ing light quanta of such short wave length that, in principle, one could use it to try to measure the positions of
electrons in atoms. Heisenberg argues that such quanta
would have to be so energetic that after each measurement the electron would be knocked out of the atom—
hence that the idea of an electron "orbit" in an atom
was meaningless since it was unmeasurable. One could
predict where the electron could be found in the atom
with the greatest probability, using the Schrödinger
wave function, but nothing more. Bohr saw in this
something even deeper—a whole new philosophical outlook which he called "complementarity," and which he
felt illuminated the limitations on the use of concepts
not only in physics but throughout science and philosophy. Just as in physics where there are pairs of complementary variables such as position and momentum
of such a character that the more precisely one is determined then, necessarily, the more imprecise must be the
determination of the other, Bohr saw in such ancient
philosophical questions as the separation of subject from
object, or the role of love versus justice, the interplay of
mutually restrictive complementary concepts.

From the beginning Einstein rejected the uncertainty
principle. As he wrote to Schrödinger in 1928, "The
Heisenberg-Bohr tranquilizing philosophy or religion?
—is so delicately contrived that, for the time being, it
provides a gentle pillow for the true believer from
which he cannot very easily be aroused. So let him lie
there."[6] As one might imagine, Einstein was not content to let "the true believers"—who, after a while, included most physicists—lie on their pillows. He began
almost at once to formulate apparent paradoxes in the
theory which Bohr responded to, one after the other,
even after Einstein's death. As people who knew Bohr
often pointed out, it was as if each day he began from
the beginning reviewing all of his arguments in real or

[6] Klein, *Letters on Wave Mechanics*, p. 31.

imaginary dialogues with Einstein to make sure he had left nothing out. In 1948 Bohr wrote a masterly summary of his discussions with Einstein over the years which he concluded by stating, "Whether our actual meetings have been of short or long duration, they have always left a deep and lasting impression on my mind, and when writing this report I have, so-to-say, been arguing with Einstein all the time even when entering on topics apparently far removed from the special problems under debate at our meetings."[7] Bohr recounted one of their most celebrated discussions, which took place at the Solvay Conference in Brussels in 1930. For this occasion Einstein had invented a remarkable imaginary device involving clocks and scales—perhaps drawing on his experience examining patents in Bern— which appeared to violate the uncertainty principle. After a sleepless night Bohr discovered that in his reasoning Einstein had forgotten to take into account the effect of *his own discovery* that clocks run at a slower rate in a gravitational field, and that, indeed, the uncertainty principle was secure.

All of the principles in this debate—which lasted nearly three decades—remained unbudgeable to the end. (Ehrenfest, who knew both men well and witnessed many of their discussions, was tormented by his own conflicting feelings as to who was right, and this, it is generally agreed, was one of the factors that led to his suicide in 1933.) Max Born summarized the attitude of many physicists when he wrote in 1948 of Einstein, "He has seen more clearly than anyone before him the statistical background of the laws of physics, and he was a pioneer in the struggle of conquering the wilderness of quantum phenomena. Yet later, when out of his own work a synthesis of statistical and quantum principles emerged which seemed acceptable to almost all physi-

[7] Niels Bohr, "Discussion with Einstein on Epistemological Problems in Atomic Physics," in Paul Arthur Schilpp, ed., *Albert Einstein: Philosopher-Scientist*, p. 240.

cists he kept himself aloof and sceptical. Many of us regard this as a tragedy—for him, as he gropes his way in loneliness, and for us, who miss our leader and standard-bearer."[8] Sometime before writing this Born had received a letter from Einstein in which he wrote, "In our scientific expectation we have grown antipodes. You believe in a dice-playing God and I in perfect laws in the world of things existing as real objects, which I try to grasp in a wildly speculative way."[9]

[8] Max Born, "Einstein's Statistical Theories," in Schilpp, *op. cit.*, pp. 163–64.
[9] *Ibid.*, p. 176.

Uranium and the Queen of Belgium

XVIII

October 1933 Einstein had settled in Princeton. Here, for the next twenty-two years, he
spent the winters working, often alone and sometimes with young assistants, on his attempt at
a unified field theory; and spending the sum-
mers working, playing chamber music with his
neighbors, and sailing his boat from such places
as his rented summer home in Peconic. There
were occasional visitors in the summer and
among them, in 1937, was C. P. Snow, who has
given a vivid impression of his visit:

> At close quarters, Einstein's head was as I
> had imagined it: magnificent, with a great
> humanizing touch of the comic. Great fur-
> rowed forehead; aureole of white hair; enor-
> mous bulging chocolate eyes. I can't guess
> what I should have expected from such a face
> if I hadn't known. A shrewd Swiss once said

it had the brightness of a good artisan's countenance, that he looked like a reliable old-fashioned watchmaker in a small town who perhaps collected butterflies on a Sunday.

What did surprise me was his physique. He had come in from sailing and was wearing nothing but a pair of shorts. It was a massive body, very heavily muscled: he was running to fat around the midriff and in the upper arms, rather like a foot-baller in middle-age, but he was an unusually strong man. He was cordial, simple, utterly unshy.[1]

Two years later, in mid-July 1939, the physicists Leo Szilard and Eugene Wigner paid Einstein a visit which, perhaps to oversimplify, resulted in a chain of events that led to the beginning of the atomic age. So much has been written about this, and Einstein has become implicated in it beyond reason, that it is perhaps worth while to attempt to clarify what actually happened.

As is well known, in 1938 the German chemists Otto Hahn and Franz Strassmann along with the Austrian physicist Lise Meitner of the Kaiser Wilhelm-Gesellschaft in Berlin "discovered" nuclear fission. In fact fission was produced as early as 1934 by Fermi, in Italy, and by Irene and Frederic Joliot-Curie in France, but this work was not correctly interpreted until 1938, by Lise Meitner and her nephew Otto Frisch, who first realized that the experiments of Hahn and Strassmann meant that they had in fact split the atomic nucleus. Lise Meitner had fled to Sweden—she was Jewish—and only learned of the results of the experiments she had been working on with Hahn and Strassmann through correspondence with Hahn. Hahn was baffled by the fact that when they had bombarded uranium by slow neutrons this yielded barium nuclei, which are about half the mass of the uranium nuclei. Where had the barium come from? She and Frisch soon realized that

[1] C. P. Snow, "On Albert Einstein," pp. 52–53.

what must have happened is that the uranium nuclei had been split in the process; i.e., "fissioned," a word they invented for what had occurred. News of the discovery was soon transmitted to Bohr in Copenhagen, where Frisch was working. By January 1939 physicists in the United States were made aware of fission after Bohr had discussed it with them at a scientific meeting in Washington, and soon after, several of them confirmed it in their own laboratories. Einstein had no involvement with any of this and, in fact, had been quite skeptical of the practical use of nuclear energy for anything. A few years before he is reported to have said that to him it seemed as remote as "shooting birds in the dark in a country where there are only a few birds."[2] What worried Wigner and Szilard was their realization that if Germany was going to try to build a bomb she would need large quantities of uranium and that, in fact, after Hitler's seizure of Czechoslovakia in the spring of 1939 one of the first actions taken by the Germans was to block the export of uranium from the Czech mines—a sure indication that the Germans were aware of its importance. Szilard knew that Einstein had had a close friendship with the Belgian royal family and that he was in periodic correspondence with Queen Elizabeth. It was their idea—his and Wigner's—to inform Einstein about the uranium situation so that he could write to the queen, as Belgium controlled the very important uranium deposits in the Belgian Congo, to keep them from falling into German hands and to open up an important source of supply for the United States. This was the subject of the conversation with Einstein in July when it was decided that a letter would be drafted to the queen and sent to her after having it cleared through the State Department.

[2] Otto Nathan and Heinz Norden, eds., *Einstein on Peace*, p. 290.

In the meanwhile Szilard had become concerned because the financial resources of the physics department at Columbia University, where he and Fermi were working, were not sufficient to support the research they were doing on fission. Szilard was advised to get in touch with Alexander Sachs, an economist and banker, who was an advisor to Roosevelt, with the idea of, perhaps, obtaining some federal support for the research. It was Sachs, it appears, who first realized the magnitude of the problem and suggested that it be brought to Roosevelt's personal attention. Shortly afterward Szilard and Edward Teller, who was then a visiting professor at Columbia, went back to Long Island, where Einstein dictated a first draft in German of his subsequent letter to Roosevelt. By August 2, two weeks after their first visit, Szilard had prepared a modified English translation which was approved and signed by Einstein and given to Sachs to transmit to Roosevelt. The final version read:

Sir:

Some recent work by E. Fermi and L. Szilard, which has been communicated to me in manuscript, leads me to expect that the element uranium may be turned into a new and important source of energy in the immediate future. Certain aspects of the situation seem to call for watchfulness and, if necessary, quick action on the part of the Administration. I believe, therefore, that it is my duty to bring to your attention the following facts and recommendations.

In the course of the last four months it has been made probable—through the work of Joliot in France as well as Fermi and Szilard in America—that it may become possible to set up nuclear chain reactions in a large mass of uranium, by which vast amounts of power and large quantities of new radium-like elements would be generated. Now it appears almost certain that this could be achieved in the immediate future.

This new phenomenon would also lead to the construction of bombs, and it is conceivable—though much less certain—that extremely powerful bombs of a new type may thus be constructed. A single bomb of this type, carried by boat or exploded in a port, might very well destroy the whole port together with some of the surrounding territory. However, such bombs might very well prove to be too heavy for transportation by air.

The United States has only very poor ores of uranium in moderate quantities. There is some good ore in Canada and the former Czechoslovakia, while the most important source of uranium is the Belgian Congo.

In view of this situation you may think it desirable to have some permanent contact maintained between the Administration and the group of physicists working on chain reactions in America. One possible way of achieving this might be for you to entrust with this task a person who has your confidence and who would perhaps serve in an unofficial capacity. His task might comprise the following:

1.) To approach Government Departments, keep them informed of further developments, and put forward recommendations for Government action, giving particular attention to the problem of securing a supply of uranium ore for the United States.

2.) To speed up experimental work which is at present being carried on within the limits of the budgets of University laboratories, by providing funds, if such funds be required, through his contacts with private persons who are willing to make contributions for this cause, and perhaps also by obtaining the cooperation of industrial laboratories which have the necessary equipment.

I understand that Germany has actually stopped the sale of uranium from the Czechoslovakian mines which she has taken over. That she should have taken such early action might perhaps be understood on the ground that the son of the German Under-Secretary of State, von Weizsäcker, is attached to the

Kaiser Wilhelm Institut in Berlin, where some of the American work on uranium is now being repeated.

Yours very truly,
A. Einstein[3]

It was not until October 11 that Sachs finally had the chance to present Einstein's letter to Roosevelt along with some additional documentary material from Szilard. On October 19 Roosevelt sent Einstein a brief note stating that he had "found this data of such import that I have convened a board consisting of the head of the Bureau of Standards and a chosen representative of the Army and Navy to thoroughly investigate the possibilities of your suggestion regarding the element of uranium."[4] Almost immediately, the Advisory Committee on Uranium was formed and one knows the rest. Einstein had some informal contact with the committee, of which he was not a member, but this ended in April 1940, when Einstein withdrew from further active collaboration. By that time the committee had been enlarged and reorganized, and Einstein, as far as anyone knows, had no further direct contact with the atomic-bomb project. (During the war Einstein functioned as a consultant for the Navy, but his work cannot have had anything to do with atomic bombs, since the Navy took no part in this research.) It is unlikely in any event that Einstein could have been of much service to a project like Los Alamos, since the work was mainly in nuclear engineering and nuclear physics, fields in which Einstein had no expertise at all. He may have guessed or been told what was going on, but he was never consulted on the Los Alamos project.

The first actual use of the atomic bomb on Japan came as a surprise to Einstein and filled him with grief.

[3] The letter is given in *Einstein on Peace*, pp. 294–96, along with an earlier draft and an account of the events leading up to its being written and delivered.
[4] *Ibid.*, p. 297.

After hearing the news of Hiroshima he said simply, *"Oh, weh!"*—"Oh, woe!" From that time until the end of his life, he devoted his time and his prestige—in letters, messages, articles, and interviews—to the cause of saving mankind from destroying itself in a nuclear holocaust. In reading these documents, even those dating back to 1945, one is struck by how tough, lucid, and "infinitely far-seeing" they are. Einstein was in no way the confused idealist he is sometimes made out to be. He was not in favor of giving away the "secret" of the atomic bomb. He recognized clearly that the only real secret had been revealed at Hiroshima when the United States proved that the atomic bomb could be made, for then it was only a matter of time before others would make it as well. His central idea was that we should use our temporary advantage, and the universal fear that the bomb had engendered in mankind, to bring about some kind of lasting world order. As he wrote in 1945, "Since I do not forsee that atomic energy will prove to be a boon within the near future, I have to say that, for the present, it is a menace. Perhaps it is well that it should be. It may intimidate the human race into bringing order to its international affairs, which, without the pressure of fear, undoubtedly would not happen."[5] He deeply resented the idea that he was somehow the "father" or "grandfather" of the atomic bomb and often said that if it had not been for the menace of Germany, which he had more reason than many men to understand, he would have done nothing to have hastened the process leading to its creation.

Toward the end of his life Einstein was in frequent ill health, and with the rise of the forces that led to McCarthyism in the United States, he felt an increasing alienation from American politics and society. In 1951 he wrote, as he had so often in the past, to the Queen

[5] *Ibid.*, p. 351.

Mother of Belgium, one of the few people with whom he shared his deeper feelings.

Dear Queen:

Your warm greeting pleased me no end and re-awakened happy memories. Eighteen harsh years, full of bitter disappointment, have gone by since then. All the more solace and cheer are derived from those few people who have remained courageous and straightforward. It is due to these few that one does not feel oneself altogether a stranger on this earth. You are one of them.

While it proved eventually possible, at an exceed-ingly heavy cost, to defeat the Germans, the dear Americans have vigorously assumed their place. Who shall bring them back to their senses? The German calamity of years ago repeats itself: people acquiesce without resistance and align themselves with the forces of evil. And one stands by, powerless.

Much as I should like to, it will probably not be given to me to see Brussels again. Because of a pecu-liar popularity which I have acquired, anything I do is likely to develop into a ridiculous comedy. This means that I have to stay close to my home and rarely leave Princeton.

I am done with fiddling. With the passage of years it has become more and more unbearable for me to listen to my own playing. I hope you have not suf-fered a similar fate. What has remained is the relent-less work on difficult scientific problems. The fascinating magic of that work will continue to my last breath.

Best wishes from your,
A. Einstein[6]

Einstein died on April 18, 1955. A few months before his death he had written to a friend:

And yet, to one bent by age death will come as a release; I feel this quite strongly now that I have

[6] *Ibid.*, p. 354.

grown old myself and have come to regard death like an old debt, at long last to be discharged. Still, instinctively one does everything possible to delay this last fulfillment. Thus is the game which nature plays with us. We may ourselves smile that we are like that, but we cannot free ourselves of the instinct to which we are all subject.[7]

Had he lived for a few more years Einstein would, perhaps, have been pleased at the reawakened interest that the present generation of physicists and astronomers have in general relativity and gravitation. Nowhere in all of science is there anything so exciting and mysterious as the discoveries of the new astrophysics. There are the pulsars, which most astronomers feel represent matter so collapsed by gravitation that only matter rich with neutrons remains, forming a densely packed small star just a few miles in diameter. There are the "black holes," gravitational fields caused by collapsing stars and so strong that no light can escape from them, a fascinating and important astrophysical speculation, not yet directly confirmed, of the theory of relativity. Then there are the mysterious signals that may be caused by gravitational radiation as stars collapse. The existence of these, too, was predicted by the general theory of relativity, but none of them had been detected at the time of Einstein's death. Most physicists have now come to believe—once again—that a study of gravitation may somehow bring us still closer to the secrets of the Old One, although in some union with the quantum theory that no one can yet quite visualize. Einstein did, in fact, absorb himself in the "fascinating magic" of his work until his "last breath." Beside his hospital bed the night he died lay the pages of an unfinished calculation on the unified field theory. He had planned to continue working on it the next morning.

[7] Ibid., p. 616.

SHORT BIBLIOGRAPHY

Principal Works by Einstein

Einstein on Peace, ed. Otto Nathan and Heinz Norden. New York: Schocken Books, 1960.

Essays on Science. New York: Philosophical Library, 1934.

Investigations on the Theory of the Brownian Movement, ed. R. Fürth. New York: Dover, 1956.

Letters on Wave Mechanics with Erwin Schrödinger, Max Planck and H. A. Lorentz, ed. K. Prizbrann, trans. M. J. Klein. New York: Philosophical Library, 1967.

Out of My Later Years. New York: Philosophical Library, 1950.

Relativity. New York: Crown, 1961. No better introduction to the theory exists; the book has gone through sixteen editions since it was first published in 1916. Shortly after it was written, Einstein told Philipp Frank he was sure it was so simple that any high-school student, like his stepdaughter, could understand it. When Einstein left the room, Professor Frank asked Miss Einstein whether this was true. She responded that she understood everything except the meaning of "coordinate system"—which, needless to say, is the pivotal technical idea.

The Meaning of Relativity. Princeton, N.J.: Princeton University Press, 1950.

The Principle of Relativity, with articles by H. A. Lorentz,

H. Weyl, H. Minkowski. New York: Dover, 1952. Most of
the important papers by Einstein on relativity are collected in this book in English translation.
   The original version of Einstein's 1905 article published
in the *Annalen der Physik* is all but unavailable. Many
physics libraries have the *Annalen* dating back to 1905,
but one finds that the "Einstein volume"—No. 17—is
mysteriously missing. In consequence most readers of
the paper do so in translations usually made by physicists.
It is very easy to fall into the trap of assuming that
footnotes in these translations are in fact Einstein's when
they are not. For example, a commonly used translation
is found in this book. Here, the footnotes were added by
the great German physicist Arnold Sommerfeld, who made
connections with the literature of the period which were
not known to Einstein in 1905. I am grateful to Professors G. Holton and A. Miller for cautioning me about this
and for sorting out Einstein's very few footnotes from the
rest.

On the Method of Theoretical Physics. The Herbert Spencer
Lecture delivered June 10, 1933. Oxford: Oxford University Press, 1933.

Further Reading

Bernstein, Jeremy. *A Comprehensible World.* New York:
Random House, 1967.
Bonnor, William. *The Mystery of the Expanding Universe.*
New York: Macmillan, 1964.
Born, Max. *The Born-Einstein Letters.* New York: Walker,
1971.
———. *Einstein's Theory of Relativity.* New York: Dover,
1962.
Boorse, Henry A. and Motz, Lloyd, eds. *The World of the
Atom.* 2 vols. New York: Basic Books, 1966. Contains an
English translation of Einstein's 1905 paper on the quan-
tum theory of light.
Carmeli, Moshe, *et. al. Proceedings of the Relativity Con-
ference in the Midwest, 1969.* New York: Plenum Press,
1970. Note particularly the article by Eugene Guth, "Con-
tribution to the History of Einstein's Geometry as a
Branch of Physics," pp. 161–207.
Eddington, Arthur. *The Mathematical Theory of Relativity.*
Cambridge, England: Cambridge University Press, 1952.
———. *Space, Time and Gravitation.* Paperback ed., New
York: Harper & Brothers, 1959.

————. *The Expanding Universe.* Paperback ed., Ann Arbor: University of Michigan Press, 1958.

Frank, Philipp. *Einstein: His Life and Times.* New York: Alfred A. Knopf, 1947. Still the most profound biographical study.

Goldberg, Stanley. "Henri Poincaré and Einstein's Theory of Relativity," *American Journal of Physics,* Vol. 35 (1967), 933–44.

————. "In Defense of Ether," in Russell McCormmach, ed. *Historical Studies in the Physical Sciences 1970.* Philadelphia: University of Pennsylvania Press, 1970, pp. 89–125.

Hall, Tord. *Carl Friedrich Gauss: A Biography,* tr. Albert Frodenborg. Cambridge, Mass.: M.I.T. Press, 1970.

Heisenberg, Werner. *Physics and Beyond: Encounters and Conversations,* New York: Harper & Row, 1971.

Holton, Gerald. "Einstein and the Crucial Experiment," *American Journal of Physics,* Vol. 37 (1969), 968–82.

————. "Einstein, Michelson, and the Crucial Experiment," *Isis,* Vol. 60 (1969), 133–197.

————. "Influences on Einstein's Early Work in Relativity Theory." *American Scholar,* Vol. 37, No. 1 (Winter 1967–68).

————. "Mach, Einstein and the Search for Reality." *Daedalus,* Spring 1968.

————. "On the Origins of the Special Theory of Relativity." *American Journal of Physics,* Vol. 31 (1963), 37–47.

Infeld, Leopold. *Albert Einstein, His Work and Influences on Our World.* New York: Charles Scribner's Sons, 1950.

Katz, Robert. *An Introduction to the Special Theory of Relativity.* Princeton: D. Van Nostrand, 1964.

Klein, Martin J. "Einstein and the Wave-Particle Duality." *The Natural Philosopher* 3 (1964), pp. 1–49.

————. "Einstein, Specific Heats, and the Early Quantum Theory." *Science,* Vol. 148 (1965).

————. "Einstein's First Paper on Quanta." *The Natural Philosopher,* 2 (1963), pp. 57–86.

————. *Paul Ehrenfest.* Amsterdam: North-Holland Publishing Co., 1970.

————. "Thermodynamics in Einstein's Thought." *Science,* Vol. 157 (1967), pp. 509–516.

Koslow, Arnold. *The Changeless Order.* New York: George Braziller, 1967.

Lorentz, Hendrick A. *The Theory of Electrons.* New York: Dover, 1952.

McCormmach, Russell. "Einstein, Lorentz and the Electron

Theory," in Russell McCormmach, ed., *Historical Studies in the Physical Sciences 1970*. Philadelphia: University of Pennsylvania Press, 1970.

MacDonald, D. K. C., *Faraday, Maxwell and Kelvin*. Paperback ed., New York: Anchor, 1964.

Manuel, Frank E. *A Portrait of Isaac Newton*. Cambridge, Mass.: Harvard University Press, 1968.

Michelson, A. A. *Studies in Optics*. Chicago: University of Chicago Press, 1927.

Millikan, Robert A. *The Electron*. Chicago: University of Chicago Press, 1917.

Moller, C. *The Theory of Relativity*. Oxford: Oxford University Press, 1952.

Moszkowski, Alexander *Conversations with Einstein*. New York: Horizon Press, 1970. While one may regret that Moszkowski did not know enough physics to ask more penetrating questions of Einstein, who was at the time of the conversations at the absolute pinnacle of his creative power, some of the exchanges are extraordinarily revealing.

Planck, Max. *Survey of Physical Theory*. New York: Dover, 1960.

Poincaré, Henri. *Mathematics and Science: Last Essays (1913)*. New York: Dover, 1963.

———. *Science and Method*. New York: Dover, 1960.

Schilpp, Paul Arthur, ed., *Albert Einstein: Philosopher-Scientist*. Evanston, Ill.: The Library of Living Philosophers Inc., 1949. The contributors to this collection, mostly physicists, represent a cross-section of the most distinguished physicists of the twentieth century. The essays concern virtually all aspects of Einstein's scientific and philosophical discoveries, and since these concern most of twentieth-century physics the essays are an almost complete education in the foundation of modern physics.

Sciama, D. W. *The Physical Foundations of General Relativity*. New York: Doubleday, 1969.

Shankland, Robert S. "Conversations with Albert Einstein." *American Journal of Physics*, Vol. 31 (1963), 37–47.

Snow, C. P. "On Albert Einstein." *Commentary*, March 1967, pp. 45–55.

Weyl, Hermann. *Philosophy of Mathematics and Natural Science*. Princeton, New Jersey: Princeton University Press, 1949.

Williams, L. Pearce, ed. *Relativity Theory: Its Origins and Impact on Modern Thought*. New York: John Wiley & Sons, 1968.

INDEX

Some other titles in the Penguin Modern Masters

This acclaimed series presents studies of men who have changed and will change the thought of our time. Although the work of these leading figures is difficult and often misunderstood, Penguin Modern Masters succeeds in delineating and clarifying it. As the series' editor, Frank Kermode, has said, "The authors of these volumes are themselves masters, and they have written their books in the belief that general discussion of their subjects will henceforth be more informed and more exciting than ever before."

ANTONIN ARTAUD
Martin Esslin

ANTONIO GRAMSCI
James Joll

CLAUDE LEVI-STRAUSS
Edmund Leach

EMILE DURKHEIM
Anthony Giddens

FERDINAND DE SAUSSURE
Jonathan Culler

FRIEDRICH ENGELS
David McLellan

JOHN MAYNARD KEYNES
D. F. Moggridge

KARL MARX
David McLellan

LEON TROTSKY
Irving Howe

MARCEL PROUST
Roger Shattuck

NOAM CHOMSKY
John Lyons